T0331182

The Science of People and Office Design

The Science of People and Office Design: Planning for Thinking, Discussing and Achieving has been written for practitioners who would like to apply evidence-based and human-centric design principles to office and workplace design. Practitioner-researcher disconnects often arise due to a lack of meaningful connection between their professional worlds. This book seeks to rectify this disconnect and make it clear that research can significantly affect the likelihood that design projects achieve the objectives outlined in their briefs, and that practitioners need to have a real influence on research conducted.

This book consists of 16 chapters, each grouped into three major sections, with an overview chapter and a conclusion drawn together by the editors. Each chapter addresses a real-world workplace design-related issue. The first part of each chapter presents the editors' overview of the problem in practice and the objectives that must be achieved via design solutions. The second portion of each chapter presents peer-reviewed research related to the chapter's topic, written by a researcher focused on practical issues. The final part of each chapter is written by a workplace design practitioner and details their efforts to resolve the section's real-world workplace design-related concern by applying relevant peer-reviewed research.

This book is aimed at professionals working in business, interior design, architecture, surveying, facilities management, building services engineering, human resources and psychology who are key stakeholders in the design and delivery of modern office spaces. Postgraduates studying design, architecture, engineering, facilities management, environmental psychology and wellbeing will also find this book useful.

Sally Augustin, PhD, is a practising environmental psychologist, a principal at Design With Science and a founder of The Space Doctors. Dr Augustin is also a fellow of the American Psychological Association, the editor of Research Design Connections and a core researcher at the Interdisciplinary Center for Healthy Workplaces at the University of California, Berkeley.

Nigel Oseland, PhD, CPsychol, is an environmental (chartered) psychologist, workplace strategist, change manager, researcher, speaker and author. Dr Oseland co-founded and hosts the Workplace Trends series of conferences; he presents internationally on workplace psychology and occasionally lectures at UCL and Liverpool John Moores University. He is the author of *A Practical Guide to Post-Occupancy Evaluation and Researching Building User Experience* and *Beyond the Workplace Zoo*, published by Routledge.

The Science of People and Office Design

Planning for Thinking, Discussing and Achieving

Edited by Sally Augustin and Nigel Oseland

Routledge
Taylor & Francis Group

LONDON AND NEW YORK

Cover image: olaser/E+ via Getty Images

First published 2025
by Routledge
4 Park Square, Milton Park, Abingdon, Oxon OX14 4RN

and by Routledge
605 Third Avenue, New York, NY 10158

Routledge is an imprint of the Taylor & Francis Group, an informa business

British Library Cataloguing-in-Publication Data
A catalogue record for this book is available from the British Library

ISBN: 978-1-032-48802-8 (hbk)
ISBN: 978-1-032-47844-9 (pbk)
ISBN: 978-1-003-39084-8 (ebk)

DOI: 10.1201/9781003390848

Typeset in Times New Roman
by codeMantra

Contents

16   To work!   239

SALLY AUGUSTIN AND NIGEL OSELAND

About the editors

Sally Augustin, PhD, is a practising environmental psychologist, a principal at Design With Science, and a founder of The Space Doctors. She has extensive experience integrating neuroscience-based insights to develop recommendations for the design of places, objects and services that support desired cognitive, emotional and physical outcomes/experiences. Her client base is worldwide and includes organisations and individuals that produce and/or use designed solutions. Augustin, who is a fellow of the American Psychological Association, is the editor of Research Design Connections, which reports and synthesises (in everyday language) the findings of recent and classic research in neuroscience, cognitive science, and the social sciences that are useful to designers.

Dr Augustin, is also a core researcher at the Interdisciplinary Center for Healthy Workplaces at the University of California, Berkeley. Dr Augustin's work has been discussed in publications such as The New York Times and *The Wall Street Journal* and she holds leadership positions in professional organisations such as the Transdisciplinary Workplace Research Network, the Academy of Neuroscience for Architecture, and the Environmental Design Research Association. She has discussed using design to enhance lives on mass-market national television and radio programs in multiple countries. Dr Augustin is the author of *Designology* (2019), *Place Advantage: Applied Psychology for Interior Architecture* (2009) and, with Cindy Coleman, *The Designer's Guide to Doing Research: Applying Knowledge to Inform Design* (2012).

Nigel Oseland, PhD, CPsychol, is an environmental (chartered) psychologist, workplace strategist, change manager, researcher, speaker and published author with 11 years research and 26 years workplace consulting experience. Through his consulting practice, Workplace Unlimited, Dr Oseland draws on his psychology background and his own research to advise occupiers on how to redefine their workstyles and rethink their workplace to create working environments that enhance individual and organisational performance, delivering maximum benefit. He specialises in strategic briefing and change management to help create workplaces that meet psychological needs and facilitate collaboration, creativity and concentration. Dr Oseland has advised corporate businesses, public sector bodies, educational institutions and professional bodies in the UK and throughout EMEA.

Through his research and consulting career, Dr Oseland has written over 100 papers and numerous workplace guides for BCO, BRE, CIBSE, IWFM and other UK institutions. He has published three books: *Beyond the Workplace Zoo: Humanising the Office (2022)*, *A Practical Guide to Post-Occupancy Evaluation – And Researching Building User Experience* (2024) and, with Paul Bartlett, *Improving Office Productivity: A Guide for Business and Facilities Managers* (1999). Dr Oseland co-founded and hosts the Workplace Trends series of conferences; he presents internationally on workplace psychology and occasionally lectures at UCL and Liverpool John Moores University.

Contributors

Christina Bodin Danielsson, PhD, associate professor in architecture (human-environment interaction), holds a Master of Architecture degree from Lund university and doctoral degree in Architecture from KTH (the Royal Institute of Technology), Sweden. Dr Bodin Danielsson works both as researcher and practicing architect specialised on the physical environment´s impact (office, school and health environments, urban planning). This work includes design work, but also advisors to organisations, property owners and public authorities, including the European Union. She holds the position as senior researcher at both Chalmers University of Technology and KTH, currently involved in different hybrid office projects investigating its impact at different level (employee, organisational and urban planning level). A special focus of her work is on e.g. health, job satisfaction, identification, social support, conflicts.

William Browning, BED, MSRED, Hon. AIA, LEED AP, is the managing partner at Terrapin Bright Green, an environmental strategies research and consulting firm. His clients include Disney, New Songdo City, Lucasfilm, Google, Marriott, Bank of America, Salesforce, the Inn of the Anasazi, the White House and the Sydney 2000 Olympic Village. Bill was a founding member of the USGBC Board of Directors. He began research in human productivity and green buildings in the 1990s at Rocky Mountain Institute and is co-author of *Greening the Building and the Bottom Line* (1994), *The Economics of Biophilia* (2012, and 2023), *14 Patterns of Biophilic Design* (2104), *Human Spaces 2.0 Biophilic Design in Hospitality* (2017) and *Nature Inside* (2020). His work has been featured in *The Wall Street Journal*, *The New York Times*, the *Washington Post, Elle Popular Science*, and in segments by NPR, Reuters, CNN, and PBS.

Ellen Bruce Keable is a workplace strategist, social scientist and certified change management professional focused on optimising human experiences, performance and wellbeing in the built and natural worlds.

Deborah Bucci, PhD, based on the idea that leadership requires us to be whole and resilient to engage with our fullest energy, Dr Bucci has based her long-term practice of working with others to create pathways to achieve results and live a healthy, rewarding life – providing practices, tools and strategies for others to live well and lead well. She combines a lifelong interest in health and fitness with her commitment to learning and teaching to nurture growth and transformation. Her research also builds on the same values as her practice, with a focus on how research can also bring about wellbeing. She is the Principal for LiveWell Strategies. Dr Bucci received her nursing degree from Owens Technical College, her BS in Human Development from The Ohio State University, her MA in Health and Exercise Science from Furman University, MA and PhD in Organization Development and Change from Fielding Graduate University.

Susanne Colenberg, PhD, is a postdoctoral researcher at the Department of Built Environment, Aalto University, Finland. She has a background in social psychology and interior architecture academic and holds a doctorate in human-centred design research from Delft University of Technology, The Netherlands. Before re-joining academia, she worked a researcher and consultant and taught environmental psychology at the University of Amsterdam. Her current research focuses on the impact of interior office design on employee well-being, particularly their social interactions, relationships and belongingness at work.

Emily Dunn, WELL Faculty, WELL AP, Prosci-certified, is a seasoned workplace strategist, thought leader and wellbeing consultant, boasting a robust background in organisational strategy and human resources. Driven by a passion for enhancing the human experience, Emily excels in fostering meaningful discussions that align workplace goals with data-driven metrics and business outcomes. Her expertise extends to the design of the built environment, where she crafts tailored solutions to optimise real estate and elevate the stakeholder experience. Emily brings a wealth of experience spanning numerous industries, with a particular emphasis on innovation, wellbeing and the implementation of research-based, data-driven solutions.

Emily holds a Master of Science in environmental psychology from Cornell University, where her graduate-level research focused on understanding individual needs based on tasks, personality types and architectural elements. Additionally, she earned an MBA from Vanderbilt's Owen Graduate School of Business with a concentration in Human and Organizational Performance, along with a Bachelor of Science in marketing from the University of Arizona.

Kristi Gaines, PhD, is an Associate Dean of the Graduate School and professor and chair in the Department of Design at Texas Tech University. Through research that investigates ways the built environment can accommodate diverse populations through design, Dr Gaines has achieved international recognition as a leader in designing learning environments and other spaces for individuals with sensory sensitivities and developmental disorders. Her book, *Designing for Autism Spectrum Disorders* (2016), is recognised with awards from the four leading organisations for interior and environmental design. Her work is published and presented through numerous academic journals, conferences and keynote addresses. She is co-founder of the Texas Tech University Coalition for Natural Learning and is a member of a state-wide leadership team, created by the Texas Department of State Health Services. With more than two decades of professional design and teaching experience, Dr Gaines has served in leadership roles in industry organisations, and currently leads the OLE! Texas Designer Network and designer training.

Judith Heerwagen, PhD, is an environmental psychologist whose work focuses on evidence-based design and the impact of the built environment on occupant health, wellbeing and performance. Dr Heerwagen served as a research psychologist in the US General Service Administration's Office of Federal High Performance Buildings. Prior to joining GSA in 2009, Dr Heerwagen had her own research and consulting firm where she worked with clients to integrate social science research into design projects. Her clients included both the public and private sectors organisations. Dr Heerwagen was a senior scientist at the Pacific Northwest National Laboratory and a research faculty member at the University of Washington, College of Built Environments where she is currently an affiliate faculty member in architecture. She is co-editor of *Biophilic Design: The Theory, Science and Practice of Bringing Buildings to Life* (2008) which won the 2009 Publishers Award in architecture and urban planning. Dr Heerwagen also received the 2014 Design for Humanity Award from the American Society of Interior Designers.

David Hemming is a chartered civil engineer, and a fellow of both the Institution of Civil Engineers (ICE) and the Institute of Workplace and Facilities Management (IWFM). He has a comprehensive understanding of managing large and complex estates, especially across the Public Sector, but also across national boundaries and different legal jurisdictions. Hemming has worked in a broad range of senior roles across various sectors, including defence, government, health and higher education – most recently the Interim Director of Estates and Facilities Management, University of Sheffield. He champions the importance of a strategic and integrated approach to estates delivery, and has wide-ranging experiences across estate management, capital delivery, soft and hard facilities management, transport, logistics and all aspects of sustainability.

Paige Hodsman, MSc, PGDip, BA, GMBPsS, AAPA, is the concept developer and workplace specialist at Saint-Gobain Ecophon for the UK and IE, and utilises her diverse background in psychology, environmental management, office interior design and acoustics to create sustainable and productive workplaces. With over 20 years of experience in the field, she has worked on global projects from both the client and supplier sides. Hodsman is focused on creating awareness about the importance of mental and physical wellbeing in office environments and advocates for a research-based approach to office interior design. She is also the co-developer of the psychoacoustic profiling tool and specialises in solving office noise distraction and design issues related to health and wellbeing. She holds a Master of Science in psychological sciences, a Postgraduate Diploma in environmental management, and a Bachelor of Arts in interior design. In addition, she is a speaker and author of numerous research papers, book chapters and articles related to acoustics and health and wellbeing in the workplace.

Valerie Jardon has a passion for understanding people to create purposeful spaces and is a respected leader within the design industry. In her role as the Director of Strategy at IA Interior Architects, she leverages research and innovation to propel global corporate clients into the future of belonging by creating purposeful workspaces. With 18 years of experience that fuses her interior design degree with business savvy honed through an Executive MBA, Jardon takes an empathetic yet results-driven approach to extract emerging needs and unlock the potential of places and people in the workscape. Her holistic viewpoint pushes workplace innovation for companies through understanding their people in order to create spaces where they can thrive.

Craig Knight, PhD, is a doctor of workplace psychology. Dr Knight's combination of commercial focus and scientific application has helped realise increases in wellbeing of up to 45%, productivity of up to 32% and effective intelligence of up to 19%. He is also the Founding Director of Identity Realization Limited (IDR), a commercial organisation based in Scotland with global research links.

Christine Kohlert, PhD, has researched and worked in the field of working and learning spaces of the future for more than 40 years, in particular with the interaction of space and organisation. Her main focus is on user involvement in the development process, the visualisation of change processes and spatial analysis. She is also a professor at the Mediadesign University in Munich and the University in Augsburg. Prof Kohlert worked for 12 years as a research affiliate at MIT on various research projects and led seminars on space and organisation as well as on innovation. She is the author of several books. In 2018 she published the book Space for *Creative Thinking – Design Principles for Learning and Working* together with Scott Cooper. In 2020 and 2023 she published the two volumes *Human Offices* and *Human Education*.

Jos Kraal, PhD, is assistant professor in the Department of Human-Centered Design, faculty of Industrial Design Engineering at TU Delft, the Netherlands. He holds a BSc and MSc in Human Movement Sciences, an MSc in Public Health and attained his PhD in Medical Informatics at the University of Amsterdam (the Netherlands). After working on optimising digital health interventions at the hospital (MáximaMC in Eindhoven) as a researcher and project leader, he joined TU Delft as an assistant professor. Dr Kraal is currently working on the intersect of behavioural science and human-centred design research to design and develop tools and strategies to empower and engage users to change their health behaviour.

Arnold Levin is a strategy practice area leader in Gensler's Southwest region and a Strategy Director. With over 45 years of experience in design strategy, organisational design, feasibility planning, workplace design and design research, he has worked with a wide range of clients in the US, UK and Asia, helping to solve complex organisational challenges through design strategy. He has worked on projects for a diverse range of government and private sector organisations including GSA, LA County Government, Weil Gotshal & Manges, Bloomberg,

Caterpillar, the Bill & Melinda Gates Foundation, Microsoft, Capital One, and Blue Cross Blue Shield. He is a frequent speaker at Corenet, Neocon, and IFMA World Workplace conferences in both the US and Europe. His research on organisational design, design strategy and workplace strategy has been published in the *Journal of Facilities Management, Journal of Design Management* and *WorkDesign.*

Casey Lindberg, PhD, is a teaching professor at the University of Colorado in Environmental Design. Dr Lindberg is dedicated to translating scientific, human-centred insights to design through the education of emerging designers and has committed the last decade to the scientific intersection of design and health. He holds a PhD in psychology and a Master of Architecture, and his expertise spans research methods, including physiological and environmental sensing techniques. Through his work for the international architecture firm HKS, Inc., he contributes design research and strategy to workplace, retail and campus space typologies. He has published academically in several fields, including medicine, design, psychology and environmental conditions, and his work has been covered by several international media outlets. He currently serves as research advisor to the International WELL Building Institute.

Angela Loder, PhD, is President, Greening the City, a consulting firm focused on urban planning, sustainability, nature and health. While Vice President, Research at the International WELL Building Institute, Dr Loder co-lead the development of the 2020 *Global Research Agenda on Health, Well-being and the Built Environment* with the international IWBI Research Advisory, was the lead author on the 2023 12 Competencies for measuring health and wellbeing for organisations, and managed and expanded the pre-approved survey provider program for all WELL Certified projects. She also led the development of social metrics for circularity with the World Business Council for Sustainable Development. As a research scientist, strategic planner and educator, Dr Loder brings over fifteen years of experience in interdisciplinary research and partnerships around health, wellbeing and the built and natural environment. She has published extensively on urban nature and health, and her book *Small-Scale Urban Greening: Creating Places of Health, Creativity, and Ecological Sustainability* was published in 2020.

Amy Manley is the Global Principal for Jacobs' Work + Change Strategy Team. She has created new ways of working for a wide range of both public and private organisations to align mission, vision and business drivers. Manley's experience focuses on organisational dynamics and how work, culture and work processes inform business and people performance.

Michal Matlon, MSc, is an architectural psychologist helping to create places of flourishing and meaningful work, from offices to neighbourhoods. In the past, he worked for one of the largest office developers in Europe, where he developed their user experience strategy and established an internal education program. Matlon now works as a consultant at theLivingCore in Vienna. He is a passionate public speaker and writer, advocating for the creation of humane environments. Matlon is also the co-founder of the Venetian Letter, a newsletter on human-centred, science-based architecture and urban design.

Naomi Miller, FIES, FIALD, straddles the line between design and engineering at the Pacific Northwest National Laboratory in Portland OR. As a senior scientist she applies her decades of experience in architectural lighting design in Upstate New York and San Francisco to nudge design professionals, product manufacturers, governments and end users towards more visually uplifting, environmentally responsible and healthy lighting solutions. Miller researches and reports on lighting quality issues such as visual comfort, flicker and light distribution with respect to solid-state products. Lighting is an intriguing stew of economics, human factors and physics; an essential element of productivity and comfort; and a necessary element of visual delight. In 2023 she received the CIE Waldram Gold Pin for Outstanding Contributions in Applied Illuminating Engineering and is both a fellow of the Illuminating Engineering Society and the International Association of Lighting Designers.

Cynthia Milota, NCIDQ, LEED AP, FMP, WSL, is a researcher, writer and designer. She partners with clients to formulate their unique objectives: mindful of circularity, wellness, culture, talent and success measures. Milota has held roles as a consultant and as the workplace strategist for a global financial services organisation. She has published and presented her research at academic and professional conferences, held adjunct faculty positions, served on juries and editorial review boards. She currently serves as the co-chair of the Environmental Design Research Association's (EDRA) Work Environment Network, on the International Facility Management Association's (IFMA) Workplace Evolutionaries Board and holds the Circular Economy Institute's (CEI) "Circular Built Environment Specialist" certification.

Upali Nanda, PhD, Assoc. AIA, EDAC, ACHE, is Global Sector Director, innovation and a partner at HKS. Based in Detroit, Upali has extensive experience leading research projects in design practice with a focus on the impact of design on human health and perception. Dr Nanda is Executive Director of Center for Advanced Design Research and Evaluation, or CADRE, the research arm of HKS, and teaches as the associate professor of Practice at the Taubman School of Architecture and Urban Planning at University of Michigan.

Nick Perham, PhD, is a reader in Applied Cognitive Psychology at Cardiff Metropolitan University. He obtained his undergraduate degree at the University of Wolverhampton, his MSc at the University of Plymouth and his PhD at Cardiff University. He is chair of the psychology ethics committee at Cardiff Metropolitan University, a committee member of the Cognitive Section of the British Psychological Society (BPS), a member of the Higher Education Academy and was a chair of the Welsh Branch of the BPS. His research primarily focuses on the impact of auditory distraction on performance, but he is also interested in how emotion affects cognition. His research has been published in international academic journals as well as non-academic publications such as *The Conversation*.

Marie Puybaraud, PhD, is a globally recognised thought leader on the Future of Work, helping her clients decode and reimagine the future of work. Over the last 25 years, Dr Puybaraud has been shaping and leading a track of global research on emerging themes (hybrid work, the future of work, human experience, human performance, outsourcing, future skills, performance models, resilience) and advanced research themes (future scenarios, responsible real estate, health footprint, the enterprise of the future, future fit strategies, regenerative workplace, empathic neighbourhood, employee value proposition). She authors articles in major media and her global research was referenced and published worldwide. Dr Puybaraud gives keynote presentations and webinars to international audiences, Fortune 500 companies and the Public Sector. She is a member of the Board of Trustees of the IFMA Foundation and serves on the IFMA European Board. These positions reflect her commitment to advancing the field of corporate real estate and workplace strategy, as well as her dedication to contributing to industry-wide initiatives and thought leadership.

Lena Reiß, PhD, completed her studies in architecture at the Technical University of Munich. In her dissertation she intensively investigated spatial qualities, design parameters and their influence on the wellbeing as well as the health of people. As Head of Health and Wellbeing at the consultancy Drees & Sommer SE, she advises companies from the private sector. Since April 2024, Dr Reiß has been Professor of Usercentric real estate design at Frankfurt University of Applied Sciences.

Diane Rogers is a project architect with IA Interior Architects located in San Fransico with a passion for inclusive design. She combines her expertise with a Certificate in Neuroscience for Architecture to create spaces that cater to diverse needs while championing accessibility and wellbeing in every project. Roger's approach to design ensures environments are not only aesthetically pleasing but also emotionally enriching for all.

Jennifer Veitch, PhD, is a psychological scientist with expertise in environmental psychology, and has been a Research Officer at the National Research Council of Canada since 1992. She investigates interactions of people and their physical environments. Dr Veitch is known for a model of lighting quality which has influenced recommendations and standards in North America and internationally, furthermore, she has demonstrated that better-quality lighting can both reduce energy use and improve organisational effectiveness if it is designed with individual needs in mind. Dr Veitch is a fellow of the Canadian Psychological Association, the American Psychological Association, the International Association of Applied Psychology, and the Illuminating Engineering Society (IES) and is a senior member of IEEE. In 2011 she received the Waldram Gold Pin for Applied Illuminating Engineering from the International Commission on Illumination (CIE) and in 2018 she received the IES Medal Award. She is a member of the editorial advisory boards for seven scholarly journals, including the *Journal of Environmental Psychology*. Dr Veitch serves as the President of the CIE for the 2023-2027 term.

Pawel Wargocki, PhD, is a professor at the Technical University of Denmark and ISIAQ, REHVA and ASHRAE fellow. Prof Wargocki has more than 25 years of experience in research on human requirements in indoor environments. He is best known for his seminal work demonstrating that poor indoor environmental quality affects the performance of office work and learning. Recent research includes studies on human emissions, sleep quality, the development of IEQ rating schemes, and the performance of green buildings. Prof Wargocki has collaborated with leading research institutions, universities and industrial partners worldwide, such as the National University of Singapore, Jiaotong University in Shanghai, Syracuse Center of Excellence, United Technologies, and Google. He was President and long-standing board member of ISIAQ, Vice President of the Indoor Air 2008 conference and chair of ASHRAE committees.

Foreword

Dr Marie Puybaraud, Thought Leader on the Future of Work, JLL

Work is a social experience. People's behaviours (mental, social and physical) and aspirations (social and professional) are critical in making work something we desire and thrive with. For more than 150 years, we have been trying to adjust places of work to narrow-minded specifications (space, cost and location) and too often the needs of the organisation have been given priority over those of the workers. We know today that this is fundamentally wrong. Failing to adapt a working environment to the needs of office occupants for strategic and financial reasons means ignoring fundamental human needs. There is no doubt that workplaces are a complex ecosystem that requires flexibility and elasticity to morph into physical and social models continuously adapting to the rhythm of work. We must explore the multifaceted environment of people and workplace design, reaching far into the science of work and workforce dynamics.

Researchers have proven that ignoring the needs and aspirations of people, the workers and users of a space gives us an incomplete list of things that really matter. Over the last few decades, this narrow-minded view of the world of work delivered facilities that failed to motivate people. The workplaces developed did not include spaces to truly collaborate, to drive human performance to its full, to inspire individuals and teams to innovate. Some workspaces previously designed were not healthy, desirable and aspirational. The hidden truth is beyond the financial reports.

Practitioners have been fighting to uplift workplace design and operation to meet new occupant demands without compromising their financial targets, often having to deal with less space and more requirements. This has forced those practitioners to ignore the fundamental needs of the occupants. A distorted (and twisted) view of the world of work embedded the design of offices in fairly traditional layouts and functionalities not meeting systematically the needs of the workers, and, instead, meeting organisational objectives.

The social experience we live and feel through physical design, visual aesthetics, soundscapes, scentscapes and all the tactile stimuli surrounding the working environment help structure human experience. Designed elements are influencing our behaviours, our interactions, our health and our performance at work. The years surrounding the biggest health crisis we had to deal with in modern times (i.e. the infamous COVID-19 pandemic) shifted our priorities and deconstructed the workplace and workforce models we spent years fine-tuning. However, this challenging period enabled our community of researchers and practitioners to rethink grounded theories about work and provided a unique opportunity to redesign offices with people in mind (in and out of the office). A big "reset" was happening – the public came to believe that the psychology of work truly mattered. The design world dared to support a neurodiverse workforce, found that employee engagement was not just an annual statistic, determined that air quality was not only an Asian issue, found that biophilic design was truly valued and let mental-health–related issues rise to the top – sociology and anthropology were on the agenda. This is the perfect storm to explore and reimagine the future of work.

The Science of People and Office Design: Planning for Thinking, Discussing and Achieving is coming at the right time. It is a deep dive into real-world workplace design-related issues, gathering comprehensive and detailed peer-reviewed research and applied insights through the perspectives of world-renowned researchers and practitioners living and breathing workplace issues. It is bringing theory and practice together in one powerful guide.

Acknowledgement

A book of this nature takes time, effort and coordination. We thank all our contributors for their outstanding content and for their prompt responses, making our input and the process relatively trouble-free. Our eternal gratitude goes to: Christina Bodin Danielsson, William Browning, Ellen Keable, Deborah Bucci, Sussane Colenberg, Emily Dunn, Krisit Gaines, Judith Heerwagen, David Hemming, Paige Hodsman, Valerie Jardon, Craig Knight, Christine Kohlert, Jos Kraal, Arnold Levin, Casey Lindberg, Angela Loder, Amy Manley, Michal Matlon, Naomi Miller, Cynthia Milota, Upali Nanda, Nick Perham, Marie Puybaraud, Lena Reiß, Diane Rogers, Jenniffer Veitch, Pawel Wargocki and Kay Sargent. This book is dedicated to all the workplace researchers and practitioners, new and old.

1   Science-informed office design

Sally Augustin and Nigel Oseland

About this book

Prior to the contributions of the chapter authors, let us answer a few further questions about this book and office design.

Why this book now?

Decades of research in psychology and related disciplines have identified the physical conditions that enhance human wellbeing and professional performance. Unfortunately, many of those conditions are often not found in the modern office. For example, emphasis on design aesthetics or a focus on cost and space reduction may result in workspaces that do not align with scientific findings on cognitive performance, concentration, creativity or collaboration. Due to individual psychological and physiological requirements, it is not uncommon for occupant feedback to show that lighting is too bright or too dim or the wrong colour, or that temperatures are too hot or too cold or matched with an undesirable humidity. Likewise, the workspace can clash with organisational or national culture or the personality profiles of the occupants.

Workplace designers and researchers work to resolve and better understand the same issues but often are on parallel tracks with little knowledge of the fundamental forces determining the success of their own or other's efforts. Designers and researchers can, however, work together to create workplaces where occupants are happy, healthy, wealthy (at least in spirit), and wise. Developing and implementing solutions in the challenging real-world and increasingly hybrid contexts, they share a common body of knowledge, mutual and independent skill sets, and the ability to take decisive action with different levels of information.

This book bolsters bonds between researchers and designers, providing insights that are useful to each group as they go about their work. Designers can become more familiar with the findings of useful studies, conducted in laboratories and in the field, while researchers can come to better understand the challenges of applying research in practice.

DOI: 10.1201/9781003390848-1

The insights that lead to the success of workplaces designers and researchers align.

The fundamental factors underlying the insights that designers and researchers (and that rare creature, the design-researcher) need to apply to cooperatively move forward can be summarised under the rubric of the 5Cs of effective workplace design (Augustin, 2009). Places where people work to their full potential (onsite, at home, or elsewhere) and live pleasant lives while doing so should do the following.

- *Coordinate* with the task-at-hand. For example, people who are trying to focus need to be able to concentrate without distraction and eyestrain. As the chapters to follow will indicate, wellbeing and performance are influenced by a tremendous number of sensory experiences (seen, heard, felt, smelled, and occasionally tasted) as well as additional matters such as organisational and national cultures and occupant personality profiles.
- *Comfort* occupants. When bodies and minds are comfortable, for example, when they are in a space incorporating biophilic design, wellbeing and performance are likely to be at higher levels than when they are not.
- *Communicate* directly via nonverbal signals to occupants that their contributions to organisational success are recognised, respected, and valued, along with supporting dialogues, spoken and unspoken, between those present. Furniture design, architectural design, and design-induced mood, among other factors, all influence people who choose to talk to and what they say when they do.
- *Challenge* occupants to develop in ways that they value. People generally want to grow in ways that make it more likely that they will do their jobs more effectively, as discussed in more detail in the next section.
- *Continue* in use over time, evolving gradually, not shifting dramatically, but changing in psychologically manageable stages.

Offices fundamentals

What drives the people who work in offices?

Regardless of whether people work in offices or not, they all share the same core human motivations, according to Deci, Olafsen, and Ryan (2017). The common human objectives outlined by Deci and colleagues are:

- *Competence* – We thrive when we feel skilled and able to succeed as we work.
- *Autonomy* – Humans, individually and in groups, value having a comfortable level of control over their lives, and that feeling of power can be derived at least in part by an ability to influence conditions in the physical environment.
- *Relatedness* – We need to feel connected to other human beings.

When the objectives identified by the Deci et al. are satisfied, then enriched performance and wellbeing will be nurtured. To achieve these objectives in the workplace requires that the design is consistent with the task at hand – for instance, stressors impeding performance need to be eliminated. Office occupants also need to be able to make choices for themselves about which of the onsite work options they select, for example, whether they sit or stand while working at their desk, whether the window blinds are open or closed, or if they choose to put a photo of their family or their dog on their work surface. A space supports relatedness when it enables efforts to pleasantly spend time with others, via sensory experiences that boost mood, for example.

Deci, Olafsen, and Ryan (2017) conclude that

> Anyone interested in improving the work context within an organization and thus the performance and wellness of its employees could evaluate any policy or practice being considered in terms of whether it is likely to (a) allow the employees to gain competencies and/or feel confident, (b) experience the freedom to experiment and initiate their own behaviours and not feel pressured and coerced to behave as directed, and (c) feel respect and belonging in relation to both supervisors and peers.

Although everyone found in offices do share the same core goals, they will differ from each other along parameters that will be discussed in detail in the chapters in this book. For example, they will vary in their personality profiles, cultures, and if they are neurodiverse or not. Being cognisant of commonalities as well as potential differences in mindsets will improve both research and design solutions.

Why are there offices?

Elsbach and Bechky (2007) explain the fundamental reasons why workplaces exist. They suggest three core functions, as follows:

- *Instrumental functions* – "Functions that improve the performance (e.g., efficiency, quality, creativity) and satisfaction (e.g., comfort, willingness to remain with the organization) of workers."
- *Symbolic functions* – "Functions that affect the cultures and identities of organizations, and identities and images of workers."
- *Aesthetic functions* – "Functions that affect the sensory experiences of workers, including both cognitive and emotional responses to design and décor."

Offices can augment collaboration, the performance of focused work, and creativity via, for instance, sensory inputs (discussed in the chapters that follow). They may do this within the context of an organisational and national culture, for instance. However, many additional factors have an influence on individual and organisational outcomes beyond the physical environments in which people work, for example, salary schedules, performance review processes, and how annual leave is administered. Workplace design is not magical, the sorts of organisational policies and practices noted in the last sentence can significantly influence collective performance by, for instance, creating obstacles to success that design cannot overcome or driving performance to such high levels that workplace design is nearly insignificant.

Places where people gather to work continue to be important because a number of communication channels are used simultaneously to exchange information and to truly understand what a colleague (conversation partner) is sharing. Some of these are more predictable, such as words and the inflections used when spoken, gestures, facial expressions, eye contact, gaze, and the distance chosen to stand apart from someone. In contrast, other means of communication are less obvious, such as the scents we produce – Pazzaglia (2015) reports that "Body odors ... can act as an authenticator of truth and are reliably invoked to shape social relations." The research of Gün Semin and colleagues (Semin, 2012) clearly outlines the importance of scent to communicating humans:

> They hypothesized that chemicals in bodily secretions, such as sweat, would activate similar processes in both the sender and receiver, establishing an emotional synchrony of sorts ... they contradict the common assumption that human communication occurs exclusively through language and visual cues. Rather ... chemosignals act as a medium through which people can be 'emotionally synchronized' outside of conscious awareness.

While words can be readily transmitted via video conferencing systems, other forms of communication like scents, eye contact, and subtle gestures are not, for instance.

Our review of published research, material in the chapters that follow, and in the archives of published peer-reviewed studies indicates that workplaces support individual, team, and organisational wellbeing and performance by:

- supporting solo thinking and collaboration and boosting memory function, in short, by aligning with the activities planned,
- building in opportunities for cognitive refreshment,
- being designed to elevate user comfort, for example, through biophilia,
- providing occupants with a comfortable number of choices, and enabling them to control their environment,
- encouraging bonding between colleagues and the organisations they work for,
- maintaining conditions that support both mental and physical health,
- reflecting national and organisational cultures,
- seeming familiar/understandable,
- signalling that employees are valued and their contributions to organisational success are respected,
- being cautious of over-standardisation and automation that ignore occupant input, and
- accommodating the varying personality and neurodiversity of occupants.

How do offices contribute to performance?

As mentioned, worker performance is affected by organisational factors as well as by workplace design, but how much impact does design have? Oseland et al., (2022) estimated the potential occupant performance benefits of various workplace design elements, mostly environmental conditions. Their estimations are drawn from 105 vetted research studies that use a range of subjective and objective performance metrics in numerous types of space. Their study was unique because they weighted the studies based on their relevance to offices, resulting in realistic performance estimations. The team noted the median, lower and upper quartile effects of combined design factors on different performance metrics (Table 1.1). They also noted the impact of environmental conditions, and design elements, on task performance (Table 1.2).

Table 1.1 Effect of design elements on performance metrics

Core performance metric	Median	Lower Quartile	Upper Quartile
1. Increased task performance	1.5%	0.3%	3.5%
2. Reduced absenteeism	0.1%	0.1%	0.4%
3. Reduced staff attrition and increased attraction	1.9%	0.3%	4.5%
4. Increased organisational performance	4.0%	2.4%	5.0%
5. Improved health and wellbeing	0.2%	0.1%	0.6%

Note, "Workplace design/refurbishment" was defined as "a move to a newly fitted out or recently refurbished office, including design, furniture, building services, etc." Furthermore, "the individual performance benefit of multiple workplace parameters cannot simply be added, the benefits are more likely to follow a law of diminishing return, with little benefit after considering five individual design elements."

Table 1.2 Effect of design elements on task *performance*

Workplace parameter	Median	LQ	UQ
1. Indoor Air Quality/ventilation	0.5%	0.1%	2.0%
2. Lighting/daylight	0.5%	0.1%	1.8%
3. Acoustics/noise	2.7%	1.2%	6.7%
4. Space/layout	3.9%	2.7%	5.2%
5. Temperature	0.4%	0.1%	1.8%
6. Control	1.3%	0.6%	2.6%
7. Ergonomics/furniture	5.0%	4.5%	5.4%
8. Workplace design/refurbishment	2.8%	0.9%	8.2%
9. Biophilia, including views	2.9%	1.7%	4.3%

LQ-Lower Quartile
UQ-Upper Quartile

Oseland, Tucker, and Wilson's study makes it clear that workplace design can have meaningful effects on the performance of individuals, and, by extension, the organisations that employ them.

Conclusion

In *Beyond the Workplace Zoo*, Oseland (2022) collated much of the research relevant to office design. That book presented high-level experimental findings, concluding with Oseland's design solution, which he termed the landscaped office. This book, *The Science of People and Office Design*, delves deeper into the research and solutions by bringing together the views of numerous researchers and practitioners, all in one book.

References

Augustin, A. (2009). *Place Advantage: Applied Psychology for Interior Architecture*. Hoboken: Wiley.

Deci, E., Olafsen, A. & Ryan, R. (2017). Self-determination theory in work organizations: The state of a science. Annual *Review of Organizational Psychology and Organizational Behavior, 3*, 19–43.

Elsbach, K. & Bechky, B. (2007). It's more than a desk: Working smarter through leveraged office design. *California Management Review, 49*(32), 80–101.

Oseland, N.A. (2022) *Beyond the Workplace Zoo: Humanising the Office*. Oxon: Routledge.

Oseland, N.A., Tucker, M. & Wilson, H. (2022). Developing the return on workplace investment (ROWI) tool. *Corporate Real Estate Journal, 12*(2), 185–197.

Pazzaglia, M. (2015). Body and odors: Not just molecules, after all. *Current Directions in Psychological Science, 24*(4), 329–333.

Semin, G.R. (2012). *The Knowing Nose: Chemosignals Communicate Human Emotions*. Association for Psychological Science Press Release. Retrieved from: https://www.psychologicalscience.org/news/releases/the-knowing-nose-chemosignals-communicate-human-emotions.html

2 Biophilia and biophilic design

Editors' introduction

Modern-day humans are not that dissimilar from the first Homo sapiens. Creating spaces today that reflect and respect our evolution as a species is at the core of biophilic design. Workplace design must accommodate our basic, innate, human needs, as well as facilitating work activities.

Judith Heerwagen, our environmental psychologist, addresses how biophilic design can significantly influence our physical, emotional, and mental health and performance, as the researcher and designer perspectives that follow make clear. Joye (2007), for example, reports that biophilic design has a noteworthy and beneficial effect on our emotional and cognitive wellbeing. Furthermore, biophilic design elements, such as circadian lighting, can directly influence our health (Africa et al., 2019).

William Browning, our practitioner, builds on Kellert and Calabrese's (2015) original elements of biophilic design. They include environmental features such as natural light, natural shapes and forms, natural patterns and aging, spaciousness, links to local physical conditions and cultures, and what Kellert calls "evolved human relationships to nature," such as providing use-options with prospect and refuge (explained later).

Read on to learn much more about the research supporting biophilic design and how to best implement it.

DOI: 10.1201/9781003390848-2

Researcher perspective

Judith Heerwagen and William Browning

Our innate affinity to nature

> From infancy we concentrate happily on ourselves and other organisms. We learn to distinguish life from the inanimate and move toward it like moths to a porch light. To explore an affiliate with life is a deep and complicated process in mental development. To an extent still undervalued in philosophy and religion, our existence depends on this propensity, our spirit is woven from it, hope rises on its currents.
>
> E.O. Wilson, Biophilia (1984)

If there is an evolutionary basis for biophilia, as asserted by E.O. Wilson in the quote above, then contact with nature is a basic human need: not a cultural amenity, not an individual preference, but a universal primary need. Just as we need healthy food and regular exercise to flourish, we need ongoing connections with the natural world. Fortunately, our connections to nature can be provided in a multitude of ways: through gardening, walks in a park, playing in water, watching birds outside our window, or enjoying a bouquet of flowers.

The experience of nature across evolutionary time has left its marks on our minds, our behavioural patterns, and our physiological functioning. We see the ghosts of our ancestors' experiences in what we pay attention to in the natural environment, how we respond, and what the experience means to us. We have all likely had positive experiences in the natural environment that stick with us long afterwards. These moments may be connected to the sky, water, the landscape, or just quiet moments in a garden. They may be linked to colour, patterns, or natural events – such as rainbows. Other experiences in nature may also be regular, enjoyable parts of our daily life such as gardening or going for a walk in a natural setting.

To understand the deep underpinnings of biophilia and its manifestation in today's cultural and physical landscapes, we need to go back in time to our ancestral life as mobile hunters and gatherers. We evolved as a social species and for most of our existence the natural landscape provided the resources for survival, chief among them are water, fire, large trees, vegetation, animals, shelter, and vistas to enable environmental and weather assessment. As hunters and gatherers, the environment was a source of food, information, protection, and socialisation. We used the sky as an indicator of time and weather. We used fire for cooking and protection from nocturnal predators as well as a venue for storytelling. These amenities form the basis of biophilia and biophilic design (Heerwagen and Orians, 1993).

How can we use biophilia to create spaces and experiences that support thinking, creating, and solving problems while also resting the mind?

As note by Annie Murphy-Paul in *The Extended Mind* (2022), our evolved mind is "tuned to the frequencies of the natural world, not the built world where we spend most of our time ...." The importance of fire in human evolutionary history is one of these natural factors. For early humans, fire likely extended the day, provided heat, warded off predators and insects, illuminated dark places, and facilitated cooking. Campfires also may have provided a social nexus and relaxation effects that could have enhanced prosocial behaviour.

Recent research shows that the evolution of controlled fire leads to major changes in the life of early humans as long as a million years ago (Nuwer and Bell, 2014). According to Nuwer, the daytime activities of a hunting and gathering group focused on gossip and the practicalities of finding food. The nights, with firelight, evoked a different behavioural pattern focused on imagination as well as conversation that healed rifts, aided memory, and enabled broader, more imaginative thinking. The nights also soothed over daytime tensions with singing and storytelling.

According to Lynn (2014), calmer, more tolerant people would have benefited in the social milieu via fireside interactions relative to individuals less susceptible to relaxation response. In a recent study, Lyon used a randomised crossover design that disaggregated fire's sensory properties, pre- and post-test blood pressure measures were compared among 226 adults across three studies. The method focused on viewing simulated muted-fire, fire-with-sound, and control conditions, in addition to tests for interactions with hypnotisable, absorption, and prosociality. Results indicated consistent blood pressure decreases in the fire-with-sound condition, particularly with a longer duration of stimulus, and enhancing effects of absorption and prosociality. The findings confirm that hearth and campfires induce relaxation as part of a multi-sensory, absorptive, and social experience.

These nighttime behaviours around a campfire continue today for many people, as noted by Melvin Konner in his book *The Tangled Wing* (Konner, 1982). Storytelling, planning, knowledge transmission, and social bonding are key activities of modern-day campers.

The value of nature to human health and wellbeing

Improved mood and reduced stress are the most consistent benefits of nature across research studies, regardless of whether they are controlled laboratory experiments or field studies. Furthermore, contact with nature can be purely visual or multi-sensory, active engagement (walking, running, gardening) or passive viewing (Heerwagen, 2018).

The experience of nature can also be sensory and ephemeral, such as changes in natural light across the day or firelight at night. Although these factors are less studied, they have been part of our environmental experience for thousands of years (Heerwagen and Hase, 2001).

We regularly experience these benefits of nature and understand the values intuitively. An extensive body of research reinforces our personal experiences and makes a sound case for the importance of regular connection to nature as factor in our health and wellbeing. These biophilic factors include sunlight, green space, gardening, water, vegetation (especially flowers), and trees.

Qualities and attributes of nature in biophilic design

Our fascination with nature is derived not just from natural elements but also from the qualities and attributes of natural settings that people find particularly appealing and aesthetically pleasing. The goal of biophilic design is to create places imbued with positive emotional experiences — enjoyment, pleasure, interest, fascination, and wonder — that are the precursors of human attachment to and caring for place (Kellert et al., 2008).

Although these biophilic design practices are not yet integrated into standards or guidelines, there is increasing interest in this topic, particularly as it relates to sustainability and social equity. We know from everyday experience that nature is not equitably distributed in urban environments. Those who can afford to do so live near parks, have large street trees and rich landscaping around their homes, and work in places that have design amenities.

Fortunately, there are many ways to incorporate biophilic design features throughout the urban built fabric. While living nature is always highly desirable, it is possible to design with the qualities and features of nature in mind, thereby creating a more naturally evocative space. Design imagination can create many pleasing options out of this biophilic template.

Daylight

We have known for a long time that people prefer daylight environments and that they believe daylight is better for health and psychological functioning than electric light. However, a clear delineation of the health and wellbeing benefits is relatively recent. We know now that bright daylight has medicinal properties. It entrains circadian rhythms, enhances mood, promotes neurological health, and affects alertness (Figueiro et al., 2002; Heerwagen, 1990).

Outdoor green space

Research conducted in outdoor spaces expands on the benefits discovered in laboratory settings. A study of public housing projects in Chicago by Meghan et al. (2014) found benefits from having large trees close at hand. Using behavioural observations and interviews, the researchers found that housing developments with large trees attracted people to be outdoors and, once there, they talked to their neighbours and developed stronger personal ties.

Variations on a theme

Natural elements — trees, flowers, animals, and shells — show both variation and similarity in form and appearance due to growth patterns. Humphrey (1980) refers to this phenomenon as "rhyming" and proposes that it is the basis for aesthetic appreciation — a skill that evolved for classifying and understanding sensory experience, as well as the objects and features of the environment. He writes, "beautiful structures in nature and art are those which facilitate the task of classification by presenting evidence of the taxonomic relationships between things in a way which is informative and easy to grasp."

Heraclitan motion

Nature is always on the move. Sun, clouds, water, tree leaves, and grasses — all move on their own rhythms or with the aid of wind. Katcher and Wilkins (1993) hypothesise that certain kinds of movement patterns may be associated with safety and tranquillity, while others indicate danger. Movement patterns associated with safety show "Heraclitan" motion that is a soft pattern of movement that "always changes, yet always stays the same." Examples are the movement of trees or grasses in a light breeze, aquarium fish, or the pattern of light and shade created by cumulus clouds. In contrast, movement patterns indicative of danger show erratic movement and sudden change, such as changes in light and wind associated with storms, or birds fleeing from a hawk.

Change and resilience

All natural habitats show cycles of birth, death, and regeneration. Some life-like processes, such as storms and the diurnal cycle of light, also may be said to show developmental sequences. When stressed, natural spaces show remarkable signs of resilience. Yet, often in our built environments, stress leads to the onset of deterioration (e.g., vacant and abandoned buildings) that seems inevitable and incapable of renewing itself. Resilience is affected by the web of relationships that connect the composition of species within an ecological community. Waste from one animal becomes food for another; unused space becomes a niche for a newcomer; decaying trees become resources and living spaces for a variety of plants and animals. The use of recycled elements and the natural aging of materials can create the impression of resilience in built environments (Krebs and Gainer, 1985).

In his 1865 *Introduction to Yosemite and the Mariposa Grove, a Preliminary Report*, the landscape architect Frederick Law Olmsted describes seemingly contradictory experience of the brain in nature (Olmsted, 1865):

…the enjoyment of scenery employs the mind without fatigue and yet exercises it, tranquilises it and yet enlivens it; and thus, through the influence of the mind over the body, gives the effect of refreshing rest and reinvigoration to the whole system.

Viewing nature, even a picture or video of nature, can bring on a state of soft fascination, where the prefrontal cortex quiets down. After having this experience, we have better or cognitive capacity. This is called Attention Restoration Theory (Kaplan and Kaplan, 1989; Kaplan, 1995), and we now know that it only takes 40 seconds of viewing a scene of nature for this effect to occur (Lee, 2105). This view, even when simulated on a video screen, will lower blood pressure and heart rate (Kahn et al., 2008).

The visual elements make a difference as well. Statistical fractals, mathematically repeating nested patterns with variations, occur frequently in water. The dappled sunlight under trees, the make up of a fern leaf, the structure of a snowflake, the rhythm of waves traveling into a beach, or the flames dancing in a fireplace are all examples of statistical fractals. We are so attuned to seeing these patterns that when they are used in human-designed spaces or objects, it is much easier for the brain to process, which lowers our stress. This effect is called fractal fluency (Taylor, 2006; Hägerhäll et al., 2008). Similarly, we are also visually attuned to biomorphic forms and collinear patterns like waving grass, or fur on animal (Hubel and Wiesel, 1968). The collinear pattern of wood grain is another example and may explain our strong preference for wood as material (Browning et al., 2022).

Three-dimensional space

The differing spatial conditions of natural environments elicit differing psychological and physiological responses. Prospect and refuge (Appleton, 1975) are an important pair of experiences in the natural world. Prospect is an unimpeded view through space and is a condition that lowers stress (Grahn and Stigsdotter, 2010). In a refuge condition, our backs are protected and there may be some canopy overhead. It is an important spatial experience that allows us to recharge, and when combined with prospect, the effect is strengthened. The curving path that draws us through the forest is a mystery condition (Herzog and Bryce, 2007), while the stepping stones through a pond add a pleasurable element of risk (Browning et al., 2014). And the jaw-dropping experience of emerging out of enclosed pinion, juniper forest to the rim of the Grand Canyon is a spatial experience of awe. Awe experiences reset our perspective, pause our physiology, and lead to increased prosocial behaviour (Anderson et al., 2008).

Multi-sensory effects

Natural habitats are sensory-rich and convey information to all human sensory systems, including sight, sound, touch, taste, and odour. Life-supporting processes, such as fire, water, and sun, also are experienced in multi-sensory ways. Many of our built environments shun sensory embellishment, creating instead caverns of grey and beige, as well as outdoor soundscapes that stress rather than soothe. Although most of the research in environmental aesthetics focuses on the visual environment, there is growing interest in understanding how design appeals to multiple senses.

The presence of nature sounds in the built environment can be very beneficial. Water sound is incredibly effective at masking out distracting sounds in the brain (Haapakangas et al., 2011), and selected bird song increases perceptions of safety (Sayin et al., 2015).

Both the Japanese practice of "Kansei engineering" and emotion-centred design are grounded in links between sensory perception and emotional responses to the environment and specific features, enabling us to create "effective and empathic design solutions (de Lera, 2015).

But we now live and work in environments that are often devoid of nature and its many physical, social, and emotional benefits.

Can we fully transform our environments – and our behaviour – to support interaction with nature in its many forms? How do we translate the varied experiences of nature in the work environment in that way that supports the productivity, health, and wellbeing?

Practitioner perspective

William Browning and Judith Heerwagen

Patterns of biophilic design

Pulling together all of the threads of scientific evidence, it is possible to identity different experiences of nature that can be translated into the built environment. These experiences fit into three broad categories:

- nature in the space,
- natural analogues, and
- nature of the space

Direct experiences of nature such as views of landscapes, sounds, smells, daylight, water, and breezes are examples of "nature in the space". Indirect experiences or characteristics of nature such as biomorphic forms, natural fractal patterns, or the use of natural materials make up "natural analogues". Specific spatial experiences like prospect and refuge are categorised as "nature of the space" (Cramer and Browning, 2008).

Currently there are 15 different experiences of nature that can be translated into a pattern language for biophilic design (Browning et al., 2014; Browning and Ryan, 2020). The different patterns support different psychological and physiological outcomes – some support stress reduction, others improve cognitive performance, and others enhance mood or prosocial behaviour (Figure 2.1).

Economics of biophilia

Adding biophilic design elements to a workplace (Figure 2.2) can have many of the aforementioned psychological and physiological outcomes. These benefits can translate into significant economic benefits. Studies have indicated faster healing in hospitals (Ulrich, 1984), increased learning rates (Determan et al., 2019), increased sales in retail settings (Rosenbaum et al., 2018), and drops in crime in urban areas (Kuo and Sullivan, 2001). For office spaces, these benefits can include lower absenteeism, higher productivity, and higher level of worker satisfaction (Ryan et al., 2023).

In a typical American office building, pre-pandemic, salaries and benefits made up more than 90% of the annual cost per square foot. In contrast, rent was about 10% and energy costs, an important measure of carbon impact, were typically less than 1% of the total cost per square foot. Increasing productivity in an office has enormous financial leverage (Ryan et al., 2023). This has been seen in measures such as a dramatic drop in absenteeism in bank headquarters that introduced daylight sculptures, water features, and plantings into their new facility (Romm and Browning, 1995).

	Biophilic Pattern	Stress Reduction	Cognitive Performance	Emotion, Mood & Preference
Nature in the Space	Visual Connection with Nature	Heart rate, Blood pressure, Parasympathetic system activity	Mental engagement, Attentiveness	Attitude, Neurological rumination, Motivation, Future discounting
	Non-Visual Connection with Nature	Blood pressure, Stress hormones	Cognitive performance, Creativity	Perceived mental health, Tranquility, Pain management
	Non-Rhythmic Sensory Stimuli	Heart rate, Systolic blood pressure, Sympathetic nervous system		Dwell time, Behavioral attention and exploration
	Thermal & Airflow Variability	Comfort	Task performance, Productivity	Preceived temporal and spatial pleasure (alliesthesia)
	Presence of Water	Overall stress, Heart rate, Blood pressure	Cognitive performance, Creativity	Positive emotion, Tranquility
	Dynamic & Diffuse Light	Circadian system functioning, Visual comfort	Cognitive performance Behavioral performance	Attitude, Overall happiness
	Connection w/ Natural Systems	Overall health		Perception of environment
Natural Analogues	Biomorphic Forms & Patterns	Stress recovery	Learning outcomes	View preference
	Material Connection with Nature	Heart rate variability, Comfort, Calming, Blood pressure, Stress hormones	Task performance, Creativity	Material preference
	Complexity & Order	Perceptual and physiological stress responses	Environmental navigation, Learning outcomes, Mental relaxation	View preference
Nature of the Space	Prospect	Overall stress, Perceived safety, Comfort		Visual interest, Fatigue, Irritation, Boredom
	Refuge	Perceived safety		Visual preference
	Mystery			Pleasure response, Visual preference
	Risk/Peril			Pleasure response
	Awe	Stress related symptoms		Pro-social behavior, Attitude, Overall happiness

Figure 2.1 Total 15 patterns of biophilic design and outcomes (from *Nature Inside*, Browning and Ryan, 2020).

Case studies

Sacramento Municipal Utility District

In most office settings, it is difficult to, almost impossible, do measurements of production rate. However, in a call centre, the number of calls handled per hour per employee is easy to track. The Sacramento Municipal Utility District (SMUD) has a call centre located on the second floor of a LEED Gold–certified building. There is good daylighting proved by floor-to-ceiling windows along the main façade with views to trees and an adjoining field. Each workstation has an adjustable vent as part of a raised floor with underfloor air-distribution system. The workstations were located in rows placed perpendicular to the windows. Since the workers' computer screens are mounted on the workstation, the view out the window is not in their cone of vision.

As an experiment, SMUD invested $1.00 per workstation to have the workstations moved 11° off of perpendicular. This meant that any movement outdoors would now be within the peripheral vision of the workers. We respond much quickly to movement at our side (Finlay, 1982). So the leaves rustling in the trees, a bird, or butterfly flying past will get attention. If the workers look away from their screens to the outdoors, the muscles in the eye are relaxed by shifting the eye out of the near visual focus. If they look at the scene outdoors for 40 seconds or more, the prefrontal cortex will quiet down and they will have better cognitive capacity when they shift their attention back (Lee, et al., 2015). This is an example of Attention Restoration Theory being supported by the pattern Visual Connection to Nature. At SMUD, this change led to a 6% increase in hourly call handling, and a return of approximately $3,000 per workstation (Heschong, 2003).

COOKFOX

COOKFOX is a New York-based architecture and interior design firm with a long-standing commitment to green design and biophilia. In 2006, the firm moved into the top floor of an old department store on 6th Ave in Manhattan. They implemented a design that featured medium height partitions (42 inches) between workstations so that most everyone in the office has a view to a 3,600 square foot green roof. This was one of the first green roofs installed in New York and featured multiple species of succulents and was soon colonised by native grasses, insects, and birds. The space was a good example of the patterns, Prospect, and Visual Connection to Nature. In post-occupancy evaluation, it was apparent that the view to nature had a significant role in reducing stress. The space was, however, almost completely lacking areas that provided Refuge.

When the lease came up for renewal and the firm was facing a tripled rent per square foot, COOKFOX searched for a new space with occupiable green roof space. Their new space on West 57th Street near Central Park features three terraces, all of which were planted as green roofs. Two can be occupied, the east terrace as a seating area, food and native species gardens, while the west terrace has a meeting table under a pergola, other seating areas among trees and shrubs with a view to the Hudson River.

On the interior, the office has a central circulation spine that connects the east and west terraces and creates a strong Prospect condition. As in the prior location, medium-height partitions with custom planters allow views to outdoors and Visual Connection to Nature. To address one of the issues from the prior space, several Refuge spaces are part of the design. A post-occupancy evaluation of the space indicated that there was a need for additional Refuge opportunities, so some storage areas were converted to provide that experience (Browning and Ryan, 2020).

Interface Base Camp office

Interface, an international manufacturer of floor coverings with a strong commitment to sustainability and biophilic design, is based in Atlanta, Georgia. Their
headquarter staff was spread across several different locations in the Atlanta area,
with a variety of spatial conditions ranging from fancy private offices in a suburban
office tower to open plan workstations in an old warehouse building. The company
chose a 40,000-square-foot mid-century office building as their new headquarters.
The building was rather unattractive but located in the up and coming midtown
portion of Atlanta and directly across the street from a subway station.

From the outset, the Interface team knew that the building did not have sufficient square footage to house all their people if designed with traditional assigned
workstations. Instead the strategy designed by Perkins & Will was to create a building with a variety of spatial experiences, where there are very few assigned workstations and instead the Interface staff and visitors move about the building to the
space that best fits the needs of the task at hand.

A number of biophilic design strategies were incorporated into the redesign of
the building.

The original precast concrete skin was replaced with an all-glass façade. This
could have caused both an energy issue with overheating and an issue for the sense
of enclosure. The building was therefore rapped in a plastic film that is a reversed
photograph of a local forest. From a distance, the trees and branches are quite visible, and up close the spaces between the trees are a pixelated white surface. This
introduces both Biomorphic Forms and Complexity and Order through the use of
a statistical fractal pattern. Both of these patterns have been shown to lower stress.

A Visual Connection to Nature is achieved with plantings throughout the building, on the rooftop and views to trees in a park surrounding the subway station across
the street. A Non-visual Connection to Nature is achieved by using aromatic plants
on the roof deck terrace, and the sounds of the water feature on the deck as well.

Partially revealed views from stair ways and in various parts of the building draw
people up through the spaces. This is an effective use of the Mystery pattern. Some
of the circulation areas have very clear Prospect views through the space and to
the outside. The building features a number of booths, niches, phone rooms, small
meeting rooms, and furniture pieces that provide an effective mix of Refuge spaces.

In 2019, a team from Carnegie Mellon University (CMU) conducted an extensive post-occupancy evaluation of the Interface Base Camp building. The team
found that the Interface associates and visitors were in fact moving around the
building over the course of the day to occupy the spaces that best suited their needs
at varying times. The CMU team reported the highest level of user satisfaction that
they have ever observed over the 75 buildings that they have analysed. Clearly the
biophilic design strategies were making a difference in peoples' experience of the
building (Loftness et al., 2019).

Figure 2.2 The ArtAqua corporate office features a number of biophilic design patterns, including Prospect, Visual Connection to Nature, and Presence of Water.

Need for workplaces

Humans are a social animal; we need the direct interaction that meetings over computer screens cannot capture. The chance of interactions and ongoing mentoring that occur in live space is important to the ongoing health of organisations. Biophilic design makes higher levels of wellbeing and professional performance in those offices more likely. We need offices like Interface's Base Camp with a larger range of spatial experiences, ones that effectively use biophilic design to support the satisfaction and productivity of staff.

References

Africa, J., Heerwagen, J., Loftness, V. & Ryan Balagtas, C. (2019). Biophilic design and climate change: Performance parameters for health. *Frontiers in Built Environment, 5*, Article 28, 1–5.

Anderson, C. L., Monroy, M. & Keltner, D. (2008). Awe in nature heals: Evidence from military veterans, at-risk youth, and college students. *Emotion, 18*(8), 1195–1202.

Appleton, J. (1975). *Experience of Landscape*. New York: John Wiley & Sons.

Browning, W. D. & Ryan, C. O. (2020). *Nature Inside: A Biophilic Design Guide*. London: RIBA Publishing.

Browning, W. D., Ryan, C. O. & Clancy, J. O. (2014). *14 Patterns of Biophilic Design: Improving Health & Wellbeing in the Built Environment*. New York: Terrapin Bright Green.

Browning, W. D., Ryan, C. O. & DeMarco, C. (2022). *The Nature of Wood, an Exploration of the Science on Biophilic Responses to Wood*. New York: Terrapin Bright Green, LLC.

Cramer, J. S. & Browning, W. D. (2008). Chapter 22: Transforming building practices through biophilic design. In S. R. Kellert, J. Heerwagen & M. Mador (eds.), *Biophilic Design: The Theory, Science and Practice of Bringing Buildings to Life*. New York: John Wiley & Sons.

de Lera, E. (2015). Emotion-centered-design (ECD) new approach for designing Interactions that Matter. In: Marcus, A. (eds) Design, User Experience, and Usability: *Design Discourse. Lecture Notes in Computer Science, 9186*. Cham: Springer.

Determan, J., Albright, T., Browning, W., Akers, M. A., Archibald, P., Martin-Dunlop, C. & Caruolo, V. (2019). *The Impact of Biophilic Design on Student Success*. Washington DC: AIA BRIK.

Figueiro, M., Kalsher, M., Stevenson, B. C., Heerwagen, J. H. & Kampschroer, K. (2019). Circadian-effective light and its impact on alertness in office workers. *Lighting Research and Technology, 51*(2), 171–193.

Finlay, D. (1982). Motion perception in the peripheral visual field. *Perception, 11*(4), 457–462.

Grahn, P. & Stigsdotter, U. K. (2010). The relation between perceived sensory dimensions of urban green space and stress restoration. *Landscape and Urban Planning, 94*, 264–275.

Haapakangas, A., Kankkunen, E., Hongisto, V., Virjonen, P. K. & Keskinen, E. (2011). Effects of five speech masking sounds on performance and acoustic satisfaction: Implications for open-plan offices. *ACTA Acustica United with Acustica, 97*, 641–655.

Hägerhäll, C. M., Laike, T., Taylor, R., Küller, M., Küller, R. & Martin, T. P. (2008). Investigations of human EEG response to viewing fractal patterns. *Perception, 37*, 1488–1494.

Heerwagen, J. (1990). Affective functioning, light hunger and room brightness preferences. *Environment and Behavior, 22*(5), 608–635.

Heerwagen, J. (2018). Biophilia, design and the adapted mind. In T. Beatley, C. Jones & R. Raney (eds.), *Healthy Environments/Healing Spaces*. Charlottesville: University of Virginia Press.

Heerwagen, J. & Hase, B. (2001). Building biophilia: Connecting people to nature. *Environmental Design + Construction*, March/April, 30–36.

Heerwagen, J., & Orians, G.H. (1993). Affect and Aesthetics: Humans, Habitats and Aesthetics. https://api.semanticscholar.org/CorpusID:56659659

Herzog, T. R. & Bryce, A. G. (2007). Mystery and preference in within-forest settings. *Environment and Behavior, 39*(6), 779–796.

Heschong, L. (2003). *Windows and Offices: A Study of Office Worker Performance and the Indoor Environment*. Fair Oaks, CA: California Energy Commission: Pacific Gas and Electric Company.

Hubel, D. H. & Wiesel, T. N. (1968). Receptive fields and functional architecture of monkey striate cortex. *Journal of Physiology, 195*, 215–243. Cited in Gilbert, C. D. (2014). Chapter 27: Intermediate-level visual processing and visual primitives. In E. R. Kandel et al. (eds.), *Principles of Neural Science* (5th Edition). New York: McGraw Hill.

Humphrey, N. K. (1980). Natural aesthetics. In B. Mikellides (ed.), *Architecture for People*. London: Studio-Vista.

Joye, Y. (2007). Architectural lessons from environmental psychology: The case of biophilic architecture. *Review of General Psychology, 11*(4), 305–328.

Kahn, P. H. et al. (2008). A plasma display window? The shifting baseline problem in a technologically mediated natural world. *Journal of Environmental Psychology, 28*, 192–199.

Kaplan, S. (1995). The restorative benefits of nature: Toward an integrative framework. *Journal of Environmental Psychology, 15*(3), 169–182.

Kaplan, R. & Kaplan, S. (1989). *The Experience of Nature: A Psychological Perspective.* Cambridge: Cambridge University Press.

Katcher, A. H. & Wilkins, G. G. (1993). Dialog with animals: Its nature and culture. In S. R. Kellert & E. O. Wilson (eds.), *The Biophilia Hypothesis.* Washington, DC: Island Press.

Kellert, S., Heerwagen, J. & Mador, M. (eds.) (2008) *The Theory, Science, and Practice of Bringing Buildings to Life.* New York: Wiley.

Kellert, S. & Calabrese, E. (2015). *The Practice of Biophilic Design.* www.biophilic-design.com

Konner, M. (1982). *The Tangled Wing.* New York: Macmillan.

Krebs, K. & Gainer, M. (1985). *Biocycle Municipal Recycling in Rural Regions.* Emmaus: JG Press, Inc.

Kuo, F. E. & Sullivan, W. C. (2001). Environment and crime in the inner city: Does vegetation reduce crime? *Environment and Behavior, 33*(3), 343–367.

Lee, K. E., Williams, K. J. H., Sargent, L. D., Williams, N. S. G. & Johnson, K. A. (2015). 40-second green roof views sustain attention: The role of micro-breaks in attention restoration. *Journal of Environmental Psychology, 42*, 182–189.

Loftness, V., Aziz, A. & Son, Y. (2019). *POE+M Study of Interface HQ Work Environments, Draft (v10).* Pittsburgh: Center for Building Performance and Diagnostics, Carnegie Mellon University.

Lynn, C. D. (2014). Hearth and campfire influences on arterial blood pressure defraying the costs of the social brain thru fireside relaxation. *Evolutionary Psychology, 12*(5), 983–1003.

Meghan, T., Holtan, M. T., Dieterlen, S. L. & Sullivan, W. C. (2014). Social life under cover: Tree canopy and social capital in Baltimore, Maryland. *Environment and Behavior, 47*(5), 1–24.

Murphy-Paul, A. (2021). *The Extended Mind: The Power of Thinking Outside the Brain.* New York: Harper Collins.

Nuwer, R. & Bell, D. (2014). Identifying and quantifying the threats to biodiversity in the U Minh peat swamp forests of the Mekong Delta, Vietnam. *Oryx, 48*(1), 88–94.

Olmsted, F. L. (1865). *Introduction to Yosemite and the Mariposa Grove, a Preliminary Report.* Washington DC: Report to the US Congress.

Romm, J. & Browning, W. (1995). *Greening the Building and the Bottom Line.* Snowmass, CO: Rocky Mountain Institute.

Rosenbaum, M. S., Ramirez, G. C. & Camino, J. R. (2018). A dose of nature and shopping: The restorative potential of biophilic lifestyle center designs. *Journal of Retailing and Consumer Services, 40*, 66–73.

Ryan, C., Browning, W. & Walker, D. (2023). *The Economics of Biophilia: Why Designing with Nature in Mind Makes Financial Sense* (2nd Edition). New York: Terrapin Bright Green, LLC.

Sayin, E., Krishna, A., Ardelet, C., Decre, G. B. & Goudey, A. (2015). Sound and safe: The effect of ambient sound on the perceived safety of public spaces. *International Journal of Research in Marketing, 32*(4), 343–353.

Taylor, R. (2006). Reduction of physiological stress using fractal art and architecture. *Leonardo, 39*(3), 245–251.

Ulrich, R. (1984). View through a window may influence recovery from surgery. *Science, 224*(4647), 420–421.

Wilson, E. O. (1984). *Biophilia.* Cambridge: Harvard University Press.

3 Managing the air

Editors' introduction

Delivering and maintaining good air quality in the workplace is a fundamental requirement. Occupant feedback surveys, such as Leesman (2022), reveal that less than one-half (45.6%) of office workers are satisfied with the air quality in their workplace. For many occupants, it may feel like air quality is an oversight, or after-thought, despite the input of mechanical engineers in the modelling, design, and installation of the mechanical, natural, or mixed ventilation systems.

As the two sections in this chapter make clear, air quality shouldn't be an after-thought – it has a significant effect on our mental and physical health, wellbeing, and performance. Research documents the important contributions that effectively ventilated spaces can make to organisational (and individual) success. For example, and as an appetiser to the more detailed material to come, MacNaughton and team (2015) found that ventilating an office space at double, the ASHRAE minimum rate was likely to add $40 a year per person to building expenses (data were collected in seven different cities in different climate zones in the USA) and to improve occupant performance by 8%, which, for this group of test employees, was equivalent to a $6,500 increase in productivity per employee per year. The researchers also noted lower levels of absenteeism and improved user health in buildings with augmented ventilation levels (compared to ASHRAE minimum). Similarly, Allen et al. (2016) learned via data collected over six full days that cognitive performance scores were 61% higher in US office buildings with low volatile organic compounds (VOC) concentrations (as opposed to high ones) and were particularly enhanced with high outdoor ventilation rates (101% higher than the high VOC condition). The researchers determined that when users experienced the sorts of carbon dioxide levels regularly found in indoor spaces, their cognitive performance was worse than it was at lower concentrations of carbon dioxide.

Furthermore, this chapter reports that natural ventilation, via openable windows that admit fresh, clean air, is an important tenet of biophilic design.

DOI: 10.1201/9781003390848-3

Researcher perspective

Pawel Wargocki

Introduction

Indoor environments in buildings, where people spend >80–90% of their lifetime, should create conditions conducive to work, learning, leisure, sleep, and comfortable living. Consequently, a built environment should be safe: Health hazards for building occupants resulting from poor design and construction, and inadequate operation and maintenance and performance should be minimised. Negligence and failing to take any actions required to comply with appropriately set quality criteria for the indoor environment can lead to serious indoor environmental problems that could result in substantial financial costs and other undesirable consequences for health and comfort (Altomonte et al., 2020; Woods, 1989).

Achieving high indoor environmental quality (IEQ) in the built environment requires managing four different domains: thermal, acoustic, and luminous environment, and indoor air quality (IAQ). There is a large body of evidence on how each of the mentioned domains affects occupants in buildings, but at the same time the information on their interplay and interactions is rather limited (Frontczak and Wargocki, 2011). IAQ is probably the most complicated and the least understood in terms of which components and contaminants affect human responses and at which concentrations and in which combinations. But indoor air is an important exposure pathway because 8–15 kg is inhaled a day by an average person. Indoor air contains thousands of pollutants, which have their sources indoors and outdoors. There are also pollutants present produced indoors during chemical transformations, which occur because of reactive factors such as ozone or hydroxyl radicals as well as ultraviolet light.

The definition of all components and their effects on humans seems to be an impossible task since many new chemicals are introduced every day to the market and their toxicological information is often unknown or incomplete. Of the 40,000–60,000 chemicals in commerce, only 1% has been tested for toxicity, and simply, it will be impossible to characterise all of them. Another complication is that pollutants present indoors occur at very low concentrations that are difficult to measure even with the traditionally used laboratory grade instrumentation and require very advanced and expensive instruments for identification. This is the reason why IAQ is often discussed in the form of proxies rather than the actual pollutants and their mixtures. The proxies of air quality used include among others the level of outdoor presumably clean air supplied by ventilation and the concentration of carbon dioxide in occupied spaces. Sometimes methods are discussed to achieve such as the use of low-emitting materials and local exhaust capturing pollutants at source. Seldom the discussion is focused on the levels of specific pollutants for which sufficient toxicological information is known, including particle matter and a few organic and inorganic pollutants, listed by the World Health Organisation (WHO) Air Quality Guidelines (Awbi, 2002; WHO, 2010; Zhang et al., 2022).

Besides particulate matter and predominantly organic pollutants present in a gas phase, IAQ can also be contaminated by biological pollutants, including infectious pathogens. The latter are not included in the air quality guidelines, which should be considered an important omission. Again, measuring pathogens in air is complicated and therefore proxies are used, including ventilation rates and the levels of carbon dioxide (Morawska et al., 2020, 2024).

Because no other metric exists, subjective ratings are often used to characterise IAQ. The subjective ratings include the sensation and perception of odours, and satisfaction (acceptability) of air quality and air freshness. However, interpreting these ratings can be challenging, especially as regards user perceptions, which are biased by personal factors affecting the responses as well as external factors such as thermal conditions that influence the responses independently of the actual level of pollution (Berglund et al., 1999).

The quality of the air indoors may be expressed as the extent to which human requirements are met. There are, however, quite large differences between the requirements of individuals. Some people are rather sensitive to an environmental parameter and are difficult to satisfy, whereas others are less sensitive and are easier to satisfy. To cope with these individual differences, the environmental quality can be expressed by the percentage of persons who find an environmental parameter unacceptable, i.e., are dissatisfied with it. If there are few dissatisfied, the quality of the environment is high. If there are many dissatisfied, the quality is low.

In recent years, to better characterise health effects or harm that is associated with the exposure to air pollutants, the disability-adjusted life years (DALYs) criteria have been applied; DALYs are used by WHO to determine the harm associated with disease resulting in premature disability and/or death. This approach, albeit very promising because it provides information allowing prioritisation of the pollutants that should be dealt with, can only be used at the moment at the population level and not at the specific building level (Jantunen et al., 2011; Morantes et al., 2023).

Odours and scents

The air breathed indoors may contain dust, fumes, microbes, or aerosolised toxins, and according to different sources, the pollution indoors can often be 2–100 times higher than outdoors. But the exposures to outdoor pollutants also occur indoors and these exposures have considerably longer durations than the exposures outdoors. Some studies show significant harm caused by exposure to outdoor pollutants indoors (Asikainen et al., 2016; Liu et al., 2023).

Organic pollutants are the major sensory pollutants in indoor air. A typical mixture of organic pollutants indoors can contain up to 6,000 compounds of which at least 500 are emitted by humans and at least 500 are emitted by building materials and furnishing. These pollutants stimulate the olfactory sense (sensitive to >1/2 millionth odorants) and the general chemical sense (sensitive to >100 thousand irritants). The former is responsible for the perception of odour and the latter for irritation or pungency sensation. The perceived air quality can also be stimulated by changes in the humidity and temperature (enthalpy) of the inhaled air even when the chemical composition of the air is constant and the thermal sensation for the entire body is kept neutral (Fang et al., 2004). Some harmful pollutants are not sensed by humans at all, e.g., radon or carbon monoxide, which have negative effects on health.

Impartial human observers are used to evaluate how IAQ is perceived by building occupants as no direct instruments are currently available to do so (Wargocki, 2004).

Scents and fragrances can be added to air directly or indirectly, via cleaning products or heating, ventilation, and air conditioning (HVAC) systems, for instance. They can be used to evoke pleasant sensations and also to mask the unwanted and unpleasant odours. Many of these products can create potential risks in indoor environments because they can react with oxidants and produce more toxic pollutants (Weschler, 2003).

Ventilation

IAQ is often closely associated with ventilation because ventilation is the process of replacing indoor (polluted) air with outdoor (presumably fresh and clean) air. Ventilation is frequently considered a surrogate or proxy for IAQ: many studies have assessed the effects of IAQ on health, comfort, and work performance via investigations of ventilation. The use of ventilation as a proxy for IAQ is necessitated by two facts: (1) Indoor climates are multifactorial environments and often it is not known which of the pollutants indoors are responsible for the effects observed; and (2) no other metric characterising and describing IAQ has been systematically used (Wargocki et al., 2021; Wei et al., 2015). However, ventilation may not always be a good proxy. If the load of pollutants indoors is low, ventilation rate can be lower than prescribed by standards with no negative consequences on IAQ and building occupants. Moreover, when outdoor air can be polluted, using the outdoor air for ventilation without cleaning it, e.g., through open windows, can bring outdoor pollutants indoors and consequently reduce air quality indoors and increase exposure levels to unwanted substances (Asikainen et al., 2016).

Literature reviews of the effects of ventilation on health show that the ventilation rates at or below 10 L/s per person can significantly aggravate acute non-clinical health symptoms and there are indications that increasing the ventilation rate from 10–20 L/s per person may further reduce these symptoms (Seppänen et al., 1999). Some studies point out that the lowest ventilation rate at which no health effects were observed is 6–7 L/s per person (Carrer et al., 2015), while the highest ventilation rate at which no effects were observed is 25 L/s per person (Wargocki et al., 2002); when established, these rates did not take into account controlling airborne infection. If the WHO air quality guidelines are met, proposed ventilation rates (to reduce the health effects from non-communicable diseases) should be 4 L/s person. This rate is considered a base rate that must always be met in any indoor environment (Carrer et al., 2018). The literature on the role of ventilation in airborne transmission of infectious agents shows sufficient evidence that ventilation and air distribution in buildings are associated with the spread and transmission of infectious diseases (Li et al., 2007; Morawska et al., 2021). The ventilation rates necessary to reduce associated risks vary and depend on the source strength (emitted virus) and calculation method. Often, 5–6 h^{-1} is proposed as the rate reducing significantly the risk of infection; alternatively, ventilation rates that result in CO_2 in the occupied spaces at concentrations below 800 ppm are suggested. ASHRAE Standard 241 provides methodology for determining these rates for different building types (Bahnfleth and Sherman, 2023; Wargocki, 2021).

By changing IAQ, ventilation also affects cognitive performance. The mechanisms underlying these effects are not well understood but it seems reasonable to assume that people who do not feel very well and experience headaches when air quality is poor have difficulty concentrating and thinking clearly; the same will occur when building occupants are distracted by odour, experience sensory irritation, and allergic reactions, or when there are direct toxicological effects on their bodies (Clements-Croome, 2006; Wargocki, 2021). The studies examining ventilation effects on cognitive performance indicate that doubling ventilation rate compared to the rates typical indoors (6 L/s per person or below) will improve work performance by about 1.5% (Seppänen and Fisk, 2006). This is an estimated average effect that should be considered a guiding principle when design decisions are made. In some cases, this effect can be much higher and in some much lower. Completed studies do not show systematically which contaminants are actually responsible for the observed effects on work performance.

Some studies suggest that CO_2 at the levels typically occurring indoors could cause negative effects on cognitive performance, especially when decisions need to be made. Other studies have not confirmed these results either for the same cognitive performance metrics or different ones. So, high levels of CO_2 alone should not be seen as the primary reason that poor air quality degrades cognitive performance (Fisk et al., 2019). Some studies showed that higher levels of PM2.5 could cause negative effects on cognitive performance and the gaseous pollutants emitted from building materials could also cause these effects (Laurent et al., 2021). It has also been documented that emissions from humans, so-called human bio-effluents, when elevated could contribute to reduced cognitive performance (Zhang et al., 2015). This effect can also be caused by pollutants produced on human skin because of chemical reactions on these surfaces (Salvador et al., 2019). Still, no particular pollutants have been identified as responsible for the effects found on work performance, and assuming that the effects are caused by many pollutants acting in concert, ventilation seem to be the best proxy for these effects.

There is some evidence that ventilation affects short-term sick leave probably because of effects on infections. The developed quantitative relationship suggests a 10% reduction in illness or sick leave for each doubling of the outdoor air supply rate across the range of ventilation rates measured in indoor spaces (Seppänen and Fisk, 2006).

Temperature

The thermal environment affects comfort, wellbeing (health), work performance, and learning.

Thermal comfort is affected by air temperature, mean radiant temperature (MRT), relative air velocity, water vapour pressure in ambient air, activity level (heat production in the body), and the thermal resistance of clothing (clo-value). These six variables are included in the comfort model (Fanger, 1970). The premise of the comfort model is that the body must be in thermal balance so that the rate of heat loss to the environment is equal to the rate of heat production in the body, that the mean skin temperature should be at the appropriate level of comfort, and that there is a preferred rate of sweating related to metabolic rate. The effects of the thermal environment on humans are predicted by the predicted mean vote (PMV) that predicts the mean thermal assessment of a large group of persons and the predicted percentage of dissatisfied people (PPD). The comfort model predicts well the thermal response when the person is not experiencing local discomfort. For buildings without mechanical cooling, the adaptive model of thermal comfort was developed, which relates the neutral temperature indoors to the average temperature outdoors (Humphreys et al., 2015). The basic assumption of this model is that building occupants are not passive recipients of the thermal environment but play an active role in creating their thermal environment via behavioural adjustments, such as adjusting clothing, opening windows, or rescheduling activities. Their thermal responses are also modified by acclimatisation, habituation, and expectation. The existing thermal comfort models refer to steady-state conditions.

Thermal conditions can affect the performance of work in several ways. These mechanisms include, among others, distracted attention, lowered individual energy level, and exacerbated symptoms that have a negative effect on mental work (Clements-Croome, 2006). Studies show that avoiding thermal discomfort will result in improved cognitive performance. However, at temperatures above 26–27°C, achieving thermal neutrality will not be sufficient to avoid reduced performance possibly because of underlying negative effects on physiological functions (Lan et al., 2022). Temperatures below 18°C could reduce finger dexterity. These low temperature levels are not recommended by the WHO, at least in residential environments, as they may lead to increased risk of respiratory infections. The evidence obtained so far shows that thermal conditions within buildings can reduce performance by 5–15%; staying slightly cool can support work performance (Seppänen and Fisk, 2006). As in the case of ventilation, these numbers are only suggestive of potential effects and in some cases higher effects and in some lower effects can be observed.

Relative humidity

Relative humidity (RH) is related to air's water content and the actual temperature. At typical temperatures in buildings, it can vary from as low as around 10–15% to above 90%. High humidity levels can cause condensation and moisture damage that can increase the risk of mould, which has subsequent health implications.

The risk of mould may increase even at moderate levels in the space; a key factor is humidity close to the surface where mould can build up. The critical level in this region is considered to be 80%. Low RH levels have been shown to increase the risk of discomfort related to dryness and may increase the risk of infection for some respiratory viruses. RH below 20% was found to negatively affect work performance, probably because of the effects on visual acuity (Wyon et al., 2006).

The range of RH which will not meaningfully affect thermal comfort is considered to fall between 30 and 70%; some consider the relevant range to be 40–60%. Indoor environments should be managed to avoid high humidity, but the lower level of RH is debated. Dry air will have positive effect on perceived air quality and avoid the risk of moisture damage but there are also other risks involved. Integrating available information indicates that the level of RH should stay above 30% (Arundel et al., 1986).

Interactions

Building occupants are exposed to different indoor environment factors simultaneously and their combined effect impact occupant responses. Also, other parameters that are remotely related to indoor environment can influence responses of building occupants. These parameters include building and architectural factors, social relations, as well as psychological factors such as mood, stress, and depression. Most of the work performed so far to examine the effects of indoor environment on building occupants has not accounted for these interactions, and the existing standards do not provide information on how to account for them.

Mainly effects on comfort and satisfaction have been examined; the combined effects on health and work performance have not received much attention and mainly studies were made examining interactions of two components (Frontczak and Wargocki, 2011). Some studies investigated how changes in, e.g., air quality and noise levels affect the overall perception of IEQ and it was shown that a change in the perception of an individual parameter may not produce a comparable effect on the overall satisfaction with IEQ (Frontczak and Wargocki, 2011). This means that different environmental parameters have stronger or weaker effects on perceptions of quality. In some studies, the components of IEQ were ranked to assess their importance (Frontczak and Wargocki, 2011). The results suggest that satisfaction with the thermal environment has higher importance than acoustic comfort and satisfaction with air quality on perceptions of environmental quality; visual comfort was rated as having the lowest importance. These results are biased by contextual factors and variation in the parameters used to define IEQ so generalising them is difficult. There is little support in the literature for the position that all IEQ components contribute equally to overall IEQ. A reasonable approach is to ensure that none are compromised; this approach is used in the TAIL (Thermal, Acoustic Indoor air and Luminous environments) scheme proposed for rating IEQ (Wargocki et al., 2021).

If employees are asked to rate the conditions that they think influence their work performance, the parameters defining IEQ score the highest, including temperature, air quality, noise, visual comfort, and building maintenance (Frontczak and

Wargocki, 2011). These results suggest that managing IEQ is crucial for achieving conditions conducive to work.

Costs

Considerable negative effects are expected and have been documented when the IEQ is inadequate. However, only few economic estimates of the potential under-lying costs have been made. They all document significant financial benefits from investments in high IAQ providing short returns on investments and high return rates on investments made (Sowa et al., 2022).

Reductions in productivity due to indoor air pollution were estimated in the 1980s in the USA to reach circa $60 billion. This figure is compatible with other estimations carried out in the mid-1990s, also using US data, showing that improv-ing indoor environments can create potential annual savings and productivity gains from $29 to $168 billion by reducing the costs of respiratory illnesses, the costs of asthma and allergy, the costs related to SBS [sick building syndrome] symptoms, and productivity losses unrelated to health. The Multiple exercises suggest that the economic benefits from improved IAQ can be up to 60 times higher than the investment required to achieve it when the effects on work performance are taken into account, and that investments can generally be recovered in no more than two years, while the rate of return can be up to seven times higher than the minimum acceptable interest rate (Wargocki and Djukanovic, 2005; Wyon, 1996).

These few examples suggest that the economic benefits of improving conditions indoors will outweigh the necessary costs and will have considerable impact on the quality of space-users' lives.

Practitioner perspective

David Hemming

Introduction

Rarely are buildings designed with a specified level of air quality. Yet it has been established that the indoor environment affects health, productivity, and comfort of the occupants. This should cause a concern for designers, as they lack a starting point from which to frame their design intent and deliver a building that properly supports the occupants. For many though, there is a lack of understanding of this area and how it impacts the occupants both psychologically and physiologically. Their focus is more on the physical form of the building and the regulatory require-ments, or even the economic drivers which are invariable about capital outlay rather than whole life costs (BSRIA, 2006).

Even if a whole life costs approach is considered, it does not include the actual cost of the occupants. Whilst the 1:5:200 ratio[1] once put forward by the Royal Academy of Engineering (Evans et al., 1998) has been questioned, the University of Reading has put forward an evidence-based ratio of 1:0.4:12 (Hughes et al., 2004). What both ratios are demonstrating is that for every £/€/$ invested in con-structing a building an order of magnitude more is spent on the occupants. There-fore, from a pure Return on Investment perspective, more focus should be placed on optimising the productivity of the occupant.

There are internationally recognised categorisations for IEQ. These can be found in *ISO 17772* (ISO, 2017) and in greater detail in terms of performance standards in *BS EN 16798-1* (BSI, 2019). *ISO 17772* provides for a four-tier classification where categories are related to the level of expectations the occupants might have. A normal level would be "IEQ$_{II}$ – Medium". A higher level might be selected for occupants with special needs (children, elderly, handicapped, etc.). Originally, it was stated that a lower level would not provide any health risk but might decrease comfort. However, when they were produced, they were based on existing reports and standards that had not been influenced from a pandemic created by airborne pathogens. Therefore, there is no consideration of infection control when it recom-mends ventilation rates.

Indoor air quality

Once a commitment has been made to invest in the occupants of the building, not just the fabric and the aesthetics of it, a decision must be made on the standard

---

1 1:5:200 (1 = construction cost; 5 = maintenance and building-operating costs; 200 = business-oper-ating costs) for commercial office buildings over their lifetime.

expected of the IAQ. For many years, it was challenging to find a clear and simple specification to provide to the designers that stated what levels were acceptable for the constituent components and contaminants within the indoor air. Whilst much of the literature highlights that many of the pollutants found within the indoor environment are emitted from buildings itself,[2] due consideration needs to be place on the pollutants found in the immediate vicinity of the building. The siting of a building should take into consideration the level of pollutants that will be present through the different seasons (both man-made and natural).

The WHO updated its *Global Air Quality Guidelines* in 2021. The guidelines stated they "are applicable to both outdoor and indoor environments globally. Thus, they cover all settings where people spend time". It also revised previous published thresholds where scientific evidence has now demonstrated that there is a recognisable impact to health (Table 3.1).

Table 3.1 WHO-recommended levels for classical air pollutants

Pollutant	Averaging time	2021 AQG
$PM_{2.5}$ ($\mu g/m^3$)	Annual	5
	24 hours	15
PM_{10} ($\mu g/m^3$)	Annual	15
	24 hours	45
O_3 ($\mu g/m^3$)	Peak season	60
	8 hours	100
NO_2 ($\mu g/m^3$)	Annual	10
	24 hours	25
SO_2 ($\mu g/m^3$)	24 hours	40
CO ($\mu g/m^3$)	24 hours	4

Reproduced from *WHO Global Air Quality Guidelines* (AQG)

A useful introduction to this subject can be found in the BSRIA technical guide titled *Indoor Air Quality* (BSRIA, 2022) which provides a list of the most common indoor air contaminants and their legal limits (where they exist). It covers both chemical and biological pollutants. In 2022, Defra[3] published a report on IAQ that provides a detailed exploration of the issues, highlighting standards and recommended levels for contaminants. However, it does not provide a definitive list of contaminants or the recommended lower bounds though. It does signpost readers to a number of the established accreditation schemes that cover the area of IAQ and performance specification, such as BREEAM (2023), LEED (2019), and WELL (IWBI, 2015). It also does not mention SKA (RICS, 2013), Fitwel (CDC, 2021), or NABERS (2023) which cover this area, amongst many national and international schemes.

As highlighted in the British Council of Offices (BCO, 2018) document *Wellness Matters*, "whilst there is an emerging correlation between green buildings and the

2 Its construction materials, the furniture, fittings and equipment (FFE) within it or due to human activities indoors, such as combustion of fuels for cooking or heating to the printing of documents.
3 The UK Government's Department for Environment, Food & Rural Affairs.

recorded health and wellbeing of their users, certified 'green' buildings cannot – with confidence – promise to offer enhanced health or wellbeing to their users". An understanding is required of each of these accreditation schemes and where their true focus lies. The issues of health and wellbeing are not always prioritised and therefore IAQ levels are not always provided or particularly comprehensive. Standards such as WELL, NABERS Indoor Environment, and Fitwel address these concerns and provide detailed threshold levels and the background to why those levels have been chosen.

Another organisation that was signposted within the Defra report is the Institute of Air Quality Management (IAQM, 2021), who published guidance to assist the assessment of IAQ, in terms of monitoring, modelling, and mitigation in residential and non-residential buildings (Table 3.2). Within that document, they produced a

Table 3.2 IAQ pollutants and their assessment criteria

Pollutant	Averaging time	Guideline concentration ($\mu g/m^3$)[a,b,c]
NO_2	1-hour mean	200 (WHO, 2010) (WHO, 2005)
	24-hour mean[d]	25 (WHO, 2021)
	Annual	40 (WHO, 2010), 10 (WHO, 2021)
PM_{10}	24-hour mean[d]nnual	50 (WHO, 2005), 45 (WHO, 2021)
		20 (WHO, 2005), 15 (WHO, 2021)
$PM_{2.5}$	24-hour mean[d]	25 (WHO, 2005), 15 (WHO, 2021)
	Annual	10 (WHO, 2005), 5 (WHO, 2021)
CO_2	"Occupied period"	1,000 ppm (BB101 and CIBSE)
CO	15-minute mean	100,000 (WHO, 2010)
	1-hour mean	35,000 (WHO, 2010)
	8-hour mean	10,000 (WHO, 2010)
	24-hour mean[d]	4,000 (WHO, 2021)
O_3	8-hour mean[d]	100 (WHO, 2005), 100 (WHO, 2021)
	Peak season[g]	60 (WHO, 2021)
Total VOC (TVOC)	8-hour mean	300 (HM Government, 2021)
Acrolein	N/A[e]	0.1–0.5 (WHO, 2002)[d]
Formaldehyde	30-minute mean	100 (WHO, 2010)
	Annual	10 (PHE, 2019)
Benzene	Not applicable	Carcinogen; no safe level. Excess lifetime risk of leukaemia at 1 $\mu g/m^3$ is 6×10^{-6} (WHO, 2010)
Acetaldehyde	1-hour mean	1,420 (PHE, 2019)
α-pinene	30-minute mean	45,000 (PHE, 2019)
d-limonene	30-minute mean	90,000 (PHE, 2019)
Radon (Rn)	Annual[f]	200 Bq/m^3 (PHE, Accessed Nov. 2020)

[a] Ensure reference conditions (temperature, pressure etc.) are considered.
[b] Monitoring TVOC will provide no information regarding the nature of the individual compounds present, their concentrations or possible toxicity.
[c] This is not a health-based threshold criterion but rather a trigger level for action/indicator for IAQ.
[d] Average of daily maximum 8-hour mean O_3 concentration in the six consecutive months with the highest six-month running-average O_3 concentration.
[e] Acrolein is a WHO-tolerable concentration not a guideline, based on the inhalation route, lowest observable adverse effects level (LOAEL) divided by a safety factor. The LOAEL is based on long-term exposure but an averaging period is not specified. Any monitoring should be representative of long-term (annual or greater) exposure.
[f] The 99th percentile (i.e., 3–4 exceedance days per year).
Reproduced from IAQM (2021).

list of common IAQ pollutants and the best practice guidelines on concentration levels, including which organisations set the levels. This is reproduced later in this chapter.

During 2023, British Standards developed and published BS 40102-1 (BSI, 2023) to give recommendations for the measuring, monitoring, and reporting of the health and wellbeing factors influenced by the building and the building services. This was done with a view to harmonised standardisation of the health and wellbeing performance of non-domestic buildings within the building sector.

Ventilation

The decision on the type of ventilation provided within a building is crucial to manage concentrations of potentially harmful pollutants and ensure that they are diluted and removed from a space. Effective ventilation is provided by either mechanical or natural means or a combination of both. The pandemic demonstrated the importance of being able to provide fresh air to occupants, to remove the hazard of dangerous airborne pathogens, but it did potentially allow other harmful pollutants to enter a building from outside if not properly filtered.

There is regularly a performance gap between design and operation which really needs to be closed if building owners want to get what they paid for. Greater thermal efficiency in modern buildings means that natural infiltration cannot be used to make up for any design or construction inadequacies. The correct specification, careful detailing, with accurate installation and a thorough commissioning process accompanied by diligent postoccupancy optimisation, are now of equal importance in order to achieve a satisfactory ventilation system.

The amount of ventilation required for good IAQ depends on a number of factors:

- occupant density,
- occupant activities, and
- pollutant emissions within a space.

CIBSE Guide A: Environmental design (2015) is deemed the premier reference source for designers of low energy sustainable buildings within the UK and many parts of the world. It is perceived as a guide to good practice. However, it is slightly out of date as it still refers back to *BS EN 13779:2007 Ventilation for Buildings*, which was replaced by *BS ISO 17772-1* (2017). Both talk about four classifications for IAQ and they are interchangeable between both standards. For IAQ of a high standard, a space should receive more than 15 ls^{-1}/person. The average space should receive a minimal level of 10 L/s per person.

In the UK, minimum requirements are largely set within the IEQ_{III}–IEQ_{II} range and are mainly prescribed in *Approved Document F – Ventilation* (HM Government, 2021) of the Building Regulations. For offices, they are required to ensure that outdoor air is supplied to occupiable rooms at whichever of the following will provide the higher total rate:

a   10 L/s per person,
b   1 L/s per m² floor area.

These requirements present a challenge when trying to balance up the costs between energy and good IEQ especially in the postpandemic world of hybrid working. Unless occupancy can be measured in real time and is linked to a Building Management System, which has the ability to vary supply air, then ventilation has to be designed on maximum occupancy or Net Internal Area of the space. This can be unnecessarily expensive to operate. If a risk-based approach is taken to adjust the ventilation rate on the basis of the average expected occupancy, then the building operator can expect a higher level of dissatisfaction with the IEQ during periods of higher than average occupancy. This requires decisions to be made in the design phase of a building on what is a priority to the owner/operators/occupiers.

Scents and odours

Scents and odours are a really challenging area to monitor and measure. Odour is a mix of volatile chemical compounds or a single compound that triggers a reaction in the olfactory organ, generally at very low concentrations. Within any group of people, odour-detection thresholds will vary widely. A small proportion of population will be hypersensitive, and an equal number will be insensitive to an odour. A person may be relatively insensitive to one smell and abnormally sensitive to another. Therefore, trigger levels can vary in any given population depending on the specific odour. This variation between people with a healthy sense of smell across a community correlates well with the distribution shown in Figure 3.1.

The US Environmental Protection Agency (EPA, 1992) provided guidance on four sensory characteristics that are used to describe an odorous emission:

- *Hedonic tone*. This is a judgement of the relative pleasantness or unpleasantness of an odour made by assessors in an odour panel.
- *Quality/Characteristics*. This is a qualitative attribute which is expressed in terms of "descriptors", e.g., "fruity", "almond" and "fishy". This can be of use when establishing an odour source from complainants' descriptions.
- *Concentration*. The "amount" of odour present in a sample of air. It can be expressed in terms of parts per million, parts per billion, or in mg/m³ of air for a single odorous compound. More usually a mixture of compounds is present, and the concentration of the mixture can be expressed in odour units per cubic metre (ou_E/m^3).
- *Intensity*. This is the magnitude (strength) of perception of an odour (from faint to strong). Intensity increases as concentration increases but the relationship is logarithmic. Increases or decreases in concentration of an odour do not always produce a corresponding proportional change in the odour strength as perceived by the human nose.

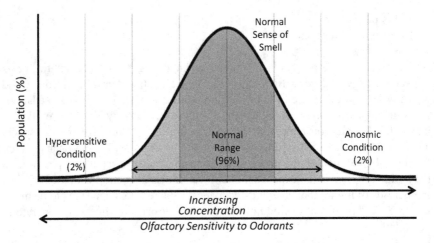

Figure 3.1 Illustration of the normal range of odour sensitivity across a potential population distribution (based on US EPA 1992; Environment Agency, 2002).

The *IPPC H4 Technical Guidance* (Environment Agency, 2002) states that a concentration of 5 ou_E/m^3 would be "faint" odour, whilst 10 ou_E/m^3 would be considered "distinct". Generally, an average person would be able to recognise the source of an odour at about 3 ou_E/m^3, although this can depend on the relative offensiveness of the odour.

There are a number of other factors which affect the variation in response to odours between individuals. These can be broadly described as:

- *Physical.* The ability to detect odours varies with age; increasing age correlates with decreasing ability. Women tend to show a slightly heightened sensitivity compared with men at any given age. Smoking habits can affect olfactory sensitivity, with smokers being less sensitive than non-smokers.
- *Psychosocial.* Once a person detects an odour, there are a number of factors which may affect the way in which he/she responds. These include the history of previous exposure, current state of health, and perception of risks to health from emissions, economic dependence on the source, expectations, coping strategies, residential satisfaction, and personality.

Also, people who are continuously exposed to a medley of "background" odours at different concentrations can often be unaware of them. Individuals may develop a "tolerance", i.e., the receptors in the nose lose sensitivity and/or the mind may screen them out.

Managing complaints about odours can be challenging as those who are dissatisfied may be small but in the hypersensitive category, so more strongly affected than the average person. It will largely be down to a pragmatic approach in an office environment, balancing costs and benefits. If the source of odour is outside of building, then looking at controlling air ingress and filtering outside air through activated carbon filters may be a suitable solution. If the source of the odour is internal, is it a temporary emission or continuous? Food preparation is a regular source of odour within an office building that has catering offer or encourages staff to bring their own food. Restricting food preparations to certain parts of the building and ensuring that ventilation is appropriate for the space is a pragmatic solution.

Temperature

The indoor thermal environment is ranked as one of the strongest contributing factors to overall human satisfaction in the built environment. Therefore, the setting of the temperature in the workplace can be the most contentious. There are long-standing recommendations from CIBSE that the optimum set point temperature for a building is 21.5°C±1.5°C. It is also recommended that room occupants have a level of control to adjust the temperature within that zone as it has been shown that occupant satisfaction is likely to be greater where there is a sense of control. Where heating or cooling is required to maintain the set point temperature of a building, it will be one of the largest single contributors to energy usage.

The thermal environment of a building not only impacts comfort and productivity levels, but also it has direct linkage to health outcomes and can impact heart rate, respiratory systems, and increase fatigue and cause negative moods. Humans can perceive temperature differently, on both physiological and psychological levels. It is more useful to talk about thermal comfort of the occupants. Thermal comfort standards utilise a model that provides a means of predicting whether humans will be satisfied with the thermal environment based on six core parameters (Figure 3.2).

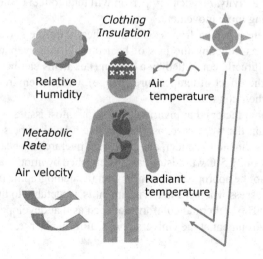

Figure 3.2 The six thermal comfort factors.

Environmental factors

Air temperature is what is normally recorded by dry bulb thermometer and is the temperature of the air that a person is in contact with. There are no legally set thresholds for minimum or maximum temperatures within the workplace. The Health and Safety Executive in the UK *advise* that temperatures should be at least 16°C or 13°C for strenuous work. These are based on background norms; different countries will take a different view on what should be acceptable.

The concept of mean radiant temperature (MRT) can be slightly harder to understand. All bodies exchange thermal radiation with their surroundings, depending on the difference in their surface temperatures and their emissivity. MRT is a measure of the average temperature of the surfaces that surround a particular point, with which it will exchange thermal radiation.

Air velocity is the velocity of the air that a person is in contact with. The faster the air is moving, the greater the exchange of heat between the person and the air. Ventilation systems look to minimise significant air movement. Section 1.3 of the current CIBSE guidance recommends that air velocities should be no more than 0.3 m/s during a heating season (CIBSE, 2021). Research has indicated that a velocity between 0.8 and 0.9 m/s has a positive and statistically significant effect on occupant's thermal comfort when cooling was needed.

Relative humidity (RH) is the ratio between the actual amount of water vapour in the air and the maximum amount of water vapour that the air can hold at that air temperature, expressed as a percentage. The higher the RH, the more difficult it is to lose heat through the evaporation of sweat.

Personal factors

Metabolic rate is the amount of energy (usually in the form of heat) we produce through physical activity. A sedentary person will tend to feel cooler than a person who is undertaking more movement.

Clothes insulate a person from exchanging heat with the surrounding air and surfaces as well as affecting the loss of heat through the evaporation of sweat. Clothing can be directly controlled by a person (i.e., items can be taken off or put on to moderate the effect of temperature change), whereas environmental factors may be beyond their control.

It is the personal factors that invariable cause the most issues when it comes to thermal comfort in the workplace, as they involve a level of personal choice and inherited genetics, linked to gender, age, race, and overarching health. The recommended set point of 21.5°C was historically established by middle-aged, Caucasian males and may not be appropriate depending on the demographic of the occupants of the space. To assess whether the environment is acceptable to the vast majority of occupants (>90%), a thermal comfort tool such as that developed by the Center for the Built Environment at the University of California Berkeley can be used.[4]

4 https://comfort.cbe.berkeley.edu/EN.

Humidity

When air temperature is within a comfortable range for occupants, the effects of humidity on thermal comfort is not greatly influenced. However, in warm temperature settings, humidity can influence the degradation of building materials; it limits the body's capacity to cool down through sweating and increases the level of discomfort from excess moisture on the skin. Outside of the direct physical effects on the occupants, warm and humid indoor spaces are also associated with mould and fungal growth. Humidity in warm spaces may promote the accumulation and growth of microbial pathogens, including bacteria and dust mites which can lead to odours and cause respiratory irritation and allergies in sensitive individuals. Conversely, cold and dry spaces can lead discomfort and irritation of the airways, skin, eyes, throat, and mucous membranes and facilitate the spread of the influenza and corona viruses.

Buildings with mechanical ventilation regularly distribute air with a RH in the 20–30% range (e.g., Edmondson, 2020). As can be seen within Figure 3.3, this is within a range that can promote many health risks to the occupants of the building. Increasing the RH to between 40 and 60% can minimise harmful effects and help reduce the spread of pathogens.

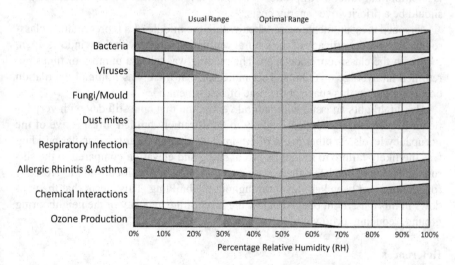

Figure 3.3 Optimum relative humidity range for minimising health risks (adapted from Sterling, Arundel & Sterling, ©ASHRAE www.ashrae.org, *ASHRAE Transactions, 91*(1), 1985)

Case study – Evaluation of an educational building for return post-pandemic

The 1,100-m² two-storey building containing multiple classrooms was completed in 2017. It was designed to be an energy-efficient building, utilising a natural ventilation system.

The ventilation strategy used a form of passive stack ventilation with classrooms on either side of a central atrium. The principle of atrium ventilation is that the air can be drawn from both sides of the building towards a central extract point, thereby effectively doubling the plan width that can be ventilated effectively by natural means. The classrooms were fed by wall-mounted louvres feeding into a ceiling plenum. Air was meant to be drawn around the classroom door which opened on the atrium and extracted through roof-mounted louvres. The air movement was meant to be aided by pressure differential between the windward and leeward sides of the building. A CO_2 monitor was placed within each classroom near the door and was set to alarm when CO_2 levels rose above a prescribed level. This alarm indicated extraction ventilation was not coping with metabolic CO_2 emission levels. The action on the monitor alarming was to open the two large top hung, bottom opening tilt windows within each classroom to increase supply ventilation. If that was not sufficient, then the classroom door was to be opened to allow greater volumes of air to be extracted through the atrium.

Whilst CO_2 levels are used as an indicator of efficiency of extraction ventilation, there is a direct correlation between CO_2 levels and cognition, as seen within the CogFX study (Cedeno Laurent et al., 2021; MacNaughton et al., 2017) undertaken by Harvard and SUNY Universities. Therefore, in spaces where cognition is an important function of productivity, CO_2 levels should be minimised, and this should be a priority over energy costs.

Airspeed measurements were taken within the middle of a representative classroom to assess the efficacy of the natural ventilation, firstly with the windows open, then with the classroom door open. The reading was taken a number of times but returned an average of 0.0m/s. This indicated that there was minimal ventilation occurring within this space at the time of assessment.

This highlights an issue with natural ventilation that on a still day with very little air movement, ventilation is likely to be extremely limited, irrespective of the occupancy levels. Another cause of minimal ventilation for this particular building was the likely failure to model its actual siting and elevation compared to the surrounding area. It had been sited on an area that had a slight slope, and therefore the foundations had been kept level throughout the building. This meant that the windows in this classroom opened below the level of the majority of the neighbouring ground, reducing airflow into the window.

References

Allen, J. G., MacNaughton, P., Satish, U., Santanam, S., Vallarino, J. and Spengler, J. D. (2016). Associations of cognitive function scores with carbon dioxide, ventilation, and volatile organic compound exposures in office workers: a controlled exposure study of green and conventional office environments. *Environmental Health Perspectives, 124*(6),805–812.

Altomonte, S., Allen, J., Bluyssen, P. M., Brager, G., Heschong, L., Loder, A., Schiavon, S., Veitch, J. A., Wang, L. & Wargocki, P. (2020). Ten questions concerning wellbeing in the built environment. *Building and Environment, 180*, 106949.

Arundel, A. V., Sterling, E. M., Biggin, J. H. & Sterling, T. D. (1986). Indirect health effects of relative humidity in indoor environments. *Environmental Health Perspectives, 65*, 351–361.

Asikainen, A., Carrer, P., Kephalopoulos, S., Fernandes, E. D. O., Wargocki, P. & Hänninen, O. (2016). Reducing burden of disease from residential indoor air exposures in Europe (HEALTHVENT project). *Environmental Health, 15*(1), 61–72.

Awbi, H. B. (2002). *Ventilation of Buildings.* London: Spon Press.

Bahnfleth, W. & Sherman, M. (2023). A first look at ASHRAE Standard 241. *ASHRAE Journal, 65*(8), 26–28.

BCO (2018). *Wellness Matters.* London: British Council of Offices.

Berglund, B., Bluyssen, P., Clausen, G., Garriga-Trillo, A., Gunnarsen, L. B., Knöppel, H., Lindvall, T., Macleod, P., Mølhave, L. & Winneke, G. (1999). *Sensory Evaluation of Indoor Air Quality.* European Collaborative Action: Indoor Air Quality and Its Impact on Man. European Commission Report, 20, 24–25.

BREEAM (2023). *BREEAM the World's Foremost Environmental Assessment Method and Rating System for Buildings.* Watford: BRE Global.

BSI (2019). *BS EN 16798-1:2019 – Energy Performance of Buildings. Ventilation for Buildings - Indoor Environmental Input Parameters for Design and Assessment of Energy Performance of Buildings Addressing Indoor Air Quality, Thermal Environment, Lighting and Acoustics.* Module M1-6. BSI Standards Publication.

BSI (2023). BS 40102-1:2023 – *Health and Well-Being and Indoor Environmental Quality in Buildings: Part 1: Health and Wellbeing in Non-Domestic Buildings: Code of Practice.* BSI Standards Publication.

BSRIA (2006). What is whole life cost analysis. *BSRIA News.* Building Services Research and Information Association. Retrieved from: https://www.bsria.com/uk/news/article/what-is-whole-life-cost-analysis/

BSRIA (2022). *TG12/2022 Indoor Air Quality.* Building Services Research and Information Association.

Carrer, P., de Oliveira Fernandes, E., Santos, H., Hänninen, O., Kephalopoulos, S. & Wargocki, P. (2018). On the development of health-based ventilation guidelines: Principles and framework. *International Journal of Environmental Research and Public Health, 15*(7), 1360.

Carrer, P., Wargocki, P., Fanetti, A., Bischof, W., Fernandes, E. D. O., Hartmann, T., Kephalopoulos, S., Palkonen, S. & Seppänen, O. (2015). What does the scientific literature tell us about the ventilation–health relationship in public and residential buildings? *Building and Environment, 94*, 273–286.

CDC (2021). *Reference Guide for the Fitwel Certification System: Workplace, v2.1 Standard.* U.S. Centers for Disease Control and Prevention and US General Services Administration.

Cedeno Laurent, J. G., MacNaughton, P., Jones, E., Young, A. S., Bliss, M., Flanigan, S., Vallarino, J., Chen, L. J., Cao, X., Allen, J. G. (2021). Associations between acute exposures to PM2.5 and carbon dioxide indoors and cognitive function in office workers: A multicountry longitudinal prospective observational study. *Environmental Research Letters*, September 9, doi: 10.1088/1748-9326/ac1bd8

CIBSE (2015). *CIBSE Guide A – Environmental Design.* Chartered Institution of Building Services Engineers.

CIBSE (2021). *CIBSE Guide A – Environmental Design (2015, Updated 2021).* Chartered Institution of Building Services Engineers.

Clements-Croome, D. (Ed.) (2006). *Creating the Productive Workplace.* Oxon: Taylor & Francis.

Defra (2022). *Indoor Air Quality.* Air Quality Expert Group report to the Department for Environment, Food and Rural Affairs. London: Crown. doi: 10.5281/zenodo.6523605.

Edmondson, C. (2020). Reducing the transmission of viruses with humidification part 1: Optimum RH range. *JMP Blog*. Greensboro: James M. Pleasants Company, Inc. Retrieved from: https://jmpcoblog.com/hvac-blog/reducing-the-transmission-of-viruses-with-humidification-part-1-optimum-rh-range

Environment Agency (2002). *Technical Guidance Note IPPC H4 – Integrated Pollution Prevention and Control (IPPC), Horizontal Guidance for Odour: Part 1 – Regulation and Permitting*. Bristol: Environment Agency.

EPA (1992). *Reference Guide to Odor Thresholds for Hazardous Air Pollutants Listed in the Clean Air Act Amendments of 1990, EPA/600/R-92/047*. Washington, D.C.: U.S. Environmental Protection Agency.

Evans, R., Haryott, R., Haste, N. & Jones, A. (1998). *The Long Term Costs of Owning and Using Buildings*. London: Royal Academy of Engineering.

Fang, L., Wyon, D. P., Clausen, G. & Fanger, P. O. (2004). Impact of indoor air temperature and humidity in an office on perceived air quality, SBS symptoms and performance. *Indoor Air, 14*(7): 74–81.

Fanger, P. O. (1970). *Thermal Comfort: Analysis and Applications in Environmental Engineering*. Copenhagen: Danish Technical Press.

Fisk, W., Wargocki, P. & Zhang, X. (2019). Do indoor CO_2 levels directly affect perceived air quality, health, or work performance? *ASHRAE Journal, 61*(9), 70–77.

Frontczak, M. & Wargocki, P. (2011). Literature survey on how different factors influence human comfort in indoor environments. *Building and Environment, 46*(4), 922–937.

HM Government (2021). *The Building Regulations 2010 Approved Document F – Ventilation, 2021 Edition for use in England*. London: RIBA Books.

Hughes, W., Ancell, D., Gruneberg, S. & Hirst, L. (2004) Exposing the myth of the 1:5:200 ratio relating initial cost, maintenance and staffing costs of office buildings. In F. Khosrowshahi, (Ed.), *20th Annual ARCOM Conference*, 1–3 September 2004, Heriot Watt University. *Association of Researchers in Construction Management, 1*, 373–381.

Humphreys, M., Nicol, F. & Roaf, S. (2015). *Adaptive Thermal Comfort: Foundations and Analysis*. London: Routledge.

IAQM (2021). *Indoor Air Quality Guidance: Assessment, Monitoring, Modelling and Mitigation, Version 1.0*. London: Institute of Air Quality Management.

ISO (2017). *ISO 17772-1:2017 Energy Performance of Buildings. Indoor Environmental Quality. Part 1: Indoor Environmental Input Parameters for the Design and Assessment of Energy Performance of Buildings*. Geneva: International Standards Organization.

IWBI (2015). *The WELL Building Standard v1*. International Well Building Institute, Delos Living LLC.

Jantunen, M., Oliveira Fernandes, E., Carrer, P. & Kephalopoulos, S. (2011). *Promoting Actions for Healthy Indoor Air (IAIAQ)*. Brussels: European Commission, Directorate-General for Health and Consumer Protection.

Lan, L., Tang, J., Wargocki, P., Wyon, D. P. & Lian, Z. (2022). Cognitive performance was reduced by higher air temperature even when thermal comfort was maintained over the 24–28 C range. *Indoor Air, 32*(1), e12916.

Laurent, J. G. C., MacNaughton, P., Jones, E., Young, A. S., Bliss, M., Flanigan, S., Vallarino, J., Chen, L. J., Cao, X. & Allen, J. G. (2021). Associations between acute exposures to PM2.5 and carbon dioxide indoors and cognitive function in office workers: A multi-country longitudinal prospective observational study. *Environmental Research Letters, 16*(9), 094047.

LEED (2019). *LEED v4.1 Building Design and Construction*. U.S. Green Building Council.

Leesman (2022). The Leesman review, issue 32, the adaptability issue. *Leesman,* London. Retrieved from: https://www.leesmanindex.com/the-leesman-review/

Li, Y., Leung, G. M., Tang, J. W., Yang, X., Chao, C. Y., Lin, J. Z., Lu, J. W., Nielsen, P. V., Niu, J., Qian, H., Sleigh, A. C., Su, H.-J. J., Sundell, J., Wong, T. W. & Yuen, P. L. (2007). Role of ventilation in airborne transmission of infectious agents in the built environment-a multidisciplinary systematic review. *Indoor Air, 17*(1), 2–18.

Liu, N., Liu, W., Deng, F., Liu, Y., Gao, X., Fang, L., Chen, Z., Tang, H., Hong, S., Pan, M., Liu, W., Huo, X., Guo, K., Ruan, F., Zhang, W., Zhao, B., Mo, J., Huang, C., Su, C., Sun, C., Zou, Z., Li, H., Sun, Y., Qian, H., Zheng, X., Zeng, X., Guo, J., Bu, Z., Mandin, C., Hänninen, O., Ji, J. S., Weschler, L. B., Kan, H., Zhao, Z. & Zhang, Y. (2023). The burden of disease attributable to indoor air pollutants in China from 2000 to 2017. *The Lancet Planetary Health, 7*(11), e900–e911.

MacNaughton, P., Pegues, J., Satish, U., Santanam, S., Spengler, J, & Allen, J. (2015). Economic, environmental and health implications of enhanced ventilation in office buildings. *International Journal of Environmental Research and Public Health, 12*(11), 14709–14722.MacNaughton, P., Satish, U., Cedeno Laurent, J. G., Flanigan, S., Vallarino, J, Coull, B., Spengler J.D. & Allen, J. G. (2017). The impact of working in a green certified building on cognitive function and health. *Building and Environment, 114*(March), 178–186.

Morantes, G., Jones, B., Molina, C. & Sherman, M. H. (2023). Harm from residential indoor air contaminants. *Environmental Science and Technology, 58*(1), 242–257.

Morawska, L., Allen, J., Bahnfleth, W., Bennett, B., Bluyssen, P. M., Boerstra, A., Buonanno, G., Cao, J., Dancer, S. J., Floto, A., Franchimon, F., Greenhaghl, T., Haworth, C., Hogeling, J., Isaxon, C., Jimenez, J. L., Kennedy, A., Kumar, P., Li, Y., Loomans, M., Marks, G., Marr, L. C., Mazzarella, L., Melikov, A., Krikor, X., Salthammer, T., Sekhar, C., Seppannen, O., Tanabe, S., Tang, J. W., Telier, R., Tham, K. W., Wargocki, P., Wierzbicka, A. & Yao, M. (2024). Mandating indoor air quality for public buildings. *Science, 383*(6690), 1418–1420.

Morawska, L., Allen, J., Bahnfleth, W., Bluyssen, P. M., Boerstra, A., Buonanno, G., Cao, J., Dancer, S., Floto, A., Franchimon, F., Greenhaghl, T., Haworth, C., Hogeling, J., Isaxon, C., Jimenez, J. L., Kurnitski, J., Li, Y., Loomans, M., Marks, G., Marr, L. C, Mazzarella, L., Melikov, A., Miller, S., Milton, D. K., Nazaroff, W., Nielsen, P. V., Noakes, C., Peccia, J., Querol, X., producers (Sekhar, C., Seppannen, O., Tanabe, S., Tang, J. W., Telier, R., Tham, K. W., Wargocki, P., Wierzbicka, A.) & Yao, M. (2021). A paradigm shift to combat indoor respiratory infection. *Science, 372*(6543), 689–691.

Morawska, L., Tang, J. W., Bahnfleth, W., Bluyssen, P. M., Boerstra, A., Buonanno, G., Cao, J., Dancer, S., Floto, A., Franchimon, F., Haworth, C., Hogeling, J., Isaxon, C., Jimenez, J. L., Kurnitski, J., Li, Y., Loomans, M., Marks, G., Marr, L. C, Mazzarella, L., Melikov, A., Miller, S., Milton, D. K., Nazaroff, W., Nielsen, P. V., Noakes, C., Peccia, J., Querol, X., Sekhar, C., Seppannen, O., Tanabe, S., Telier, R., Tham, K. W., Wargocki, P., Wierzbicka, A. & Yao, M. (2020). How can airborne transmission of COVID-19 indoors be minimised? *Environment International, 142*, 105832.

NABERS (2023). *Handbook Estimating NABERS Ratings, Version 3.0.* National Australian Built Environment Rating System. Sydney: New South Wales Government.

PHE (2019). *Indoor Air Quality Guidelines for selected Volatile Organic Compounds (VOCs) in the UK. London:* Public Health England.

RICS (2013). *Ska Rating: Good Practice Measures for Offices, Version 1.2.* London: Royal Institution of Chartered Surveyors.

Salvador, C. M., Bekö, G., Weschler, C. J., Morrison, G., Le Breton, M., Hallquist, M., Ekberg, L. & Langer, S. (2019). Indoor ozone/human chemistry and ventilation strategies. *Indoor Air, 29*(6), 913–925.

Seppänen, O. A. & Fisk, W. (2006). Some quantitative relations between indoor environmental quality and work performance or health. *HVAC&R Research, 12*(4), 957–973.

Seppänen, O. A., Fisk, W. J. & Mendell, M. J. (1999). Association of ventilation rates and CO2 concentrations with health and other responses in commercial and institutional buildings. *Indoor Air, 9*(4), 226–252.

Sowa, J., Tanabe, S. I. & Wargocki, P. (2022). Economic consequences. In *Handbook of Indoor Air Quality* (1477–1487). Singapore: Springer Nature Singapore.

Sterling, E. M., Arundel, A. & Sterling, T. D. (1985). Criteria for human exposure to humidity on occupied buildings. *ASHRAE Transactions, 91*(1), 611–622.

Wargocki, P. (2004). Sensory pollution sources in buildings. *Indoor Air, 14*(7), 82–91.

Wargocki, P. (2021). What we know and should know about ventilation. *REHVA Journal, 58*(2), 5–13.

Wargocki, P. & Djukanovic, R. (2005). Simulations of the potential revenue from investment in improved indoor air quality in an office building. *ASHRAE Transactions, 111*(2), 699–711.

Wargocki, P., Sundell, J., Bischof, W., Brundrett, G., Fanger, P. O., Gyntelberg, F., Hanssen, S. O., Harrison, P., Pickering, A., Seppänen, O. & Wouters, P. (2002). Ventilation and health in non-industrial indoor environments: report from a European multidisciplinary scientific consensus meeting (EUROVEN). *Indoor Air, 12*(2), 113–128.

Wargocki, P., Wei, W., Bendžalová, J., Espigares-Correa, C., Gerard, C., Greslou, O., Rivallain, M., Sesana, M. M., Olesen, B. W., Zirngibl, J. & Mandin, C. (2021). TAIL, a new scheme for rating indoor environmental quality in offices and hotels undergoing deep energy renovation (EU ALDREN project). *Energy and Buildings, 244*, 111029.

Wei, W., Ramalho, O. & Mandin, C. (2015). Indoor air quality requirements in green building certifications. *Building and Environment, 92*, 10–19.

Weschler, C. J. (2003). *Indoor Chemistry as a Source of Particles. Indoor Environment: Airborne Particles and Settled Dust*. Weinheim: WILEY-VCH, 167–189.

Woods, J. E. (1989). Cost avoidance and productivity in owning and operating buildings. *Occupational Medicine, 4*(4), 753–770.

WHO (2002). Concise International Chemical Assessment Document 43 – Acrolien. Geneva: World Health Organization.

WHO (2005). Air Quality Guidelines: Global Update 2005. Geneva: World Health Organization.WHO (2010). WHO Guidelines for Indoor Air Quality: Selected Pollutants. Geneva: World Health Organization. Regional Office for Europe.

WHO (2021). WHO Global Air Quality Guidelines. Geneva: World Health Organization.

Wyon, D. P. (1996, October). Indoor environmental effects on productivity. *Proceedings of IAQ, 96*, 5–15.

Wyon, D. P., Fang, L., Lagercrantz, L. & Fanger, P. O. (2006). *Experimental Determination of the Limiting Criteria for Human Exposure to Low Winter Humidity Indoors (RP-1160)*. *HVAC & Research, 12*(2), 201–213.

Zhang, Y., Hopke, P. K. & Mandin, C. (Eds.) (2022). *Handbook of Indoor Air Quality*. Singapore: Springer.

Zhang, X., Wargocki, P. & Lian, Z. (2015). Effects of exposure to carbon dioxide and human bioeffluents on cognitive performance. *Procedia Engineering, 121*, 138–142.

4 Visual aesthetics

Editors' introduction

The workplace design industry, and its clients and office occupants, spend some time talking about what workplaces look like. Not only are many humans highly attuned to visual stimuli, but also the media used to communicate about workplace design are largely visual – we see video walkthroughs of offices online, photographs are posted of existing workplaces, and renderings distributed of potential work sites. People regularly lead their discussions on where they work with narratives setting the visual scene.

This chapter focusses on visual aesthetics and the next chapter explores lighting. The topic of colour relates to both chapters and in vision-focused discussions a frequent topic that is highly debated is what colour various surfaces should be. Neuroscience research can help resolve many a rigorous debate about colour, and it will be covered in Chapter 16.

This chapter focusses on visual aesthetics in relation to beauty, art, workplace design, etc. and its impact on the wellbeing and performance of occupants, including different personality types.

DOI: 10.1201/9781003390848-4

Researcher perspective

Christina Bodin Danielsson

Introduction

Beauty and its perception are the main focus of aesthetics. Beauty is a positive aesthetic value that contrasts with ugliness as its negative counterpart (Sartwell, 2017). Visual aesthetics is a set or group of artistic elements that are perceived positively, visually stimulating, and pleasing. Thus, when this chapter discusses visual aesthetics, it refers to the beauty or pleasing appearance of things and environments from an art and physical environment perspective. With this as a starting point, this chapter is only a brief introduction to the very broad subject of visual aesthetics – a subject that spans from philosophy, environmental aesthetics (that includes art, architecture, urban design, and environmental psychology) to the fields of human-computer interaction (HCI), and neuroscience.

We perceive and understand the surrounding world with our senses and vision is only one of our senses used to do so. But in terms of environmental design, we live in a visual society, where the visual sense holds a dominant role in contemporary architecture and urban design, where less importance is given to our other senses (Ebenhard, 2009). This chapter aims to investigate the subject of visual aesthetics, its impact, and how aesthetic preference develops from an art and physical environment perspective. This approach is consistent with my professional experience as an architect.

In terms of visual aesthetic and aesthetic preference in art and architecture, there has traditionally been a fascination with the concept of Golden Ratio, used since antiquity to analyse proportions. Some suggest that this ratio appears in nature, so objects with this ratio are visually appealing. However, this subject is outside the scoop of this chapter which instead focuses on visual aesthetics from an art and physical environment perspective. The material that follows reflects the two major approaches in this field – an empirical and a philosophical approach focusing on what gives people pleasure and why, primarily concentrating on the empirical approach. My goal is hereby to provide easier access to the subject, and also to provide knowledge that is easily translated to design practice for the benefit of the end users that are visually experiencing different objects/environments. The empirical approach on which this chapter is based involves perception and cognition (Lang, 1987). Cognition is defined here as the processing of visual information, a conscious and unconscious effort to classify it (Kaplan, 1992; Ulrich, 1983).

Aesthetics preference

Aesthetics is defined as an artistic formula (i.e., principle and expression), which is developed to evoke a positive response from the observer/user of an aesthetic object and environment. The word aesthetics per se originates from Ancient Greek words meaning "sensitive," "perceptive," and "to feel" (Buckley and Jenny, 2012) and these source words describe well the core elements of aesthetics. An aesthetic experience is one that is "heightened and intensified," according to early research-ers like Dewey (1980). Yet, in contemporary theoretical models of aesthetic experience, beauty is just one of many undefined aesthetic responses (Pelowski et al., 2017).

Aesthetics preference in art and architecture concerns beauty and ugliness, and the evaluation of these factors. A special focus in this research is the differences in preference between naive and experienced viewers (i.e., between those trained in art and design professions and laymen), but the potential influence of personality traits and gender on aesthetic experiences is now frequently discussed. Research has found that people's aesthetic preference for visual stimuli from an object depends on the degree of abstraction, artistic qualities, and modification/novelty in the design of the object (Cela-Conde et al., 2002; Furnham and Walker, 2001; Hekkert and Wieringen, 1996; Neperud, 1986). Variables that affect aesthetic preference in this regard include not only psychophysical variables (e.g., luminance and contrast) but also variables like complexity and novelty in the design (Berlyne, 1970; Martindale et al., 1988). Regarding novelty versus familiarity, more recent research has found novelty more important for aesthetic appreciation, though its impact is modulated by properties of both the artwork and the beholder (Song et al., 2021).

Aesthetic preference relates not only to evaluation and appraisal (Silvia, 2005) but also to arousal (Berlyne, 1970). If and how this is reflected in brain activity has also received increased attention recently, but in this regard our knowledge is still sparse. Research has found that aesthetic preference processing involves multiple cognitive operations, which take place in different brain areas and in different time frames (Nadal et al., 2008). We do not yet have a detailed understanding of how this processing occurs, however, or if different areas of the brain are active when looking at objects perceived as aesthetically pleasing or ugly (Kawabata and Zeki, 2004). Experiments do show that the neural activities of judging beauty in abstract patterns correlate with those of judging symmetry (Jacobsen et al., 2006). Yet knowledge remains sparse about how neural activity relates to aesthetic preferences and interacts with the amygdala, the centre of emotions in the brain. For example, a review study of three neuroimaging studies on aesthetic processes and cognition by Nadal et al. (2008) did not find any significant activity in the amygdala.

Visual aesthetics – Studied within different domains

Aesthetic pleasure and human preference are mainly studied in the research fields of environmental aesthetics and of HCI and in those fields within the three domains of: (1) visual aesthetics in arts (paintings, sculptures), (2) physical environments, and (3) digital interfaces, i.e., human interactions with computers. In the following paragraphs, the two former domains will be discussed jointly due to their interrelationship.

Visual aesthetics of art and of environmental design

Scientific interest in aesthetic preference goes back to the late 19th and early 20th centuries, focusing on experimental studies about aesthetic preference to identify valid relationships between the concepts of "beauty" and "good taste" in art and architecture. Researchers also evaluated the potential impact on preferences of background factors like age, gender, educational background, as well as mood and personality traits (Gifford, 1980).

Environmental aesthetics focuses on aesthetic preferences in the physical environment and how this may influence our senses in a pleasurable way (Carlson, 2000; Naser, 1988). It includes aesthetic appreciation of built and natural environments and also of works of art situated in different interior and exterior environments (e.g., urban settings or natural environments). In each case, the philosophical challenge is the same: to determine the factors influencing aesthetic preference and how they are processed cognitively. Environmental aesthetics is the crossing point between the objective, physical characteristics of the human environment, and the subjective responses to it, as the "merging of two areas of inquiry: empirical aesthetics and environmental psychology" (Nasar, 1992).

A special interest in visual aesthetics concerns differences in preference between people trained in art and design professions and laymen. In terms of artwork, it has been found that artists favour more complexity (Berlyne, 1970). The same is true of architects who prefer higher levels of uncertainty and complexity in novel buildings (Devlin and Naser, 1989). In fact, architects and non-architects interpret and appreciate buildings differently – at the levels of buildings generally and also the specific design elements within them, e.g., glass cladding, colour uniformity (e.g., Gifford et al., 2000; Groat, 1982). These differences aside, there are exterior design features that are positively assessed by both groups, i.e., architects and laymen. They are both concerned with factors such as personalisation of facades and open spaces and quality of materials and execution, as well as with consistency with the typical architectural design of a geographical area (Da Luz Reis and Dias Lay, 2010). Consistency in facade colouring also seems to influence appreciation, with buildings in harmonious colours rated higher in preference and considered more congruent (O'Connor, 2006).

Visual aesthetic in the digital interface

Today, the digital environment is central in people's daily lives, not least in their working life. Consequently, the research interest in the subject of visual aesthetics preferences in digital interfaces has also increased (i.e., via software programs and homepages). One area that has caught interest in the domain of visual aesthetics concerns the association between personality and preferences in visual aesthetics. One such study, based on the "Big Five" personality traits (see later chapter on personality), has found that in terms of digital interface, only one of the five traits – Agreeableness – has a significant correlation with both classical and expressive aesthetics but that significant relationship is weak (Salimun et al.,

2021). According to the researchers, this result conforms with other findings that personality traits rather than aesthetic dimensions influence preference for different design features. However, other studies that investigated aesthetic preference in other domains have found people who score high on Agreeableness have stronger preferences for classical and representational art, and less for abstract art (Furnham and Avison, 1997). It has also been found that people high in the trait of Extraversion prefer high colour contrasts and bold or sharp-edged shapes in web graphics (Karsvall, 2002). Reinecke and colleagues have in studies (Reinecke and Gajos, 2014) on visual appeal found complexity in design to be more important than colourfulness as a predictor of appeal. The latter study did not focus on personality types. Regarding preference for colourfulness, differences between genders and educational background have also been found, with a lower preference among men than women, and among those with higher education (Reinecke and Gajos, 2014; Salimun et al., 2021). Furthermore, Reinecke and Gajos (2014) also found variations between people at different geographical locations with lower preference for colourfulness in Northern European countries, than in Southern Europe and Asia.

To summarise, despite this interest in the association between personality and aesthetic preference, such association is not at all clear as other factors also influence preference. Thus, personality is not a good predictor for aesthetic preference in design features.

Visual aesthetics in relation to design and environmental characteristics

A central question in visual aesthetics relates to users' perception of environmental quality and architectural design. The field of aesthetics research deals with perception of environmental indoor environmental characteristics and of specific design features, and their potential influences on office occupants' mental wellbeing, cognitive performance, productivity, and satisfaction.

Regarding visual aesthetics and environmental factors, a recent review study by Meng et al. (2023) describes the research on visual quality in decoration as both contradictory and difficult to overview. This is especially true of visual, nature-based decorations. One study finds exposure to nature posters to significantly affect anger and stress among men (Stone and English, 1998), while another finds marginal effects of viewing nature posters on mood and attention (Stone and English, 1998). Also, research on "naturalness," whether it concerns a view of nature or biophilic architecture, is the study of a design philosophy that aims to connect people with nature; it has no specific focus on visual aesthetics, but rather on cognition and health and wellbeing (Yina et al., 2018; Zhong et al., 2022).

Perception of design features and environmental characteristics depend on the physical attributes of those features (Moore and Golledge, 1976). Perceived order both in works of art and groups of buildings increases user satisfaction with visual aesthetics (Nasar, 1997). This effect is potentially related to a conscious and unconscious effort to classify visual information (Kaplan, 1992; Ulrich, 1983). Regarding preference for other design features, research is somewhat contradicting. Studies have found preferences for design features to differ between professional designers and laymen (Gifford et al., 2000; Nasar, 1997). For example, a preference for square windows was identified among architects and for circle or oval-shaped windows among laymen (Ghomeshi and Mohd Jusan, 2013).

Architects also seem to prefer fewer materials, more concrete, and white colouring, as well as simpler forms and a non-symmetric order, e.g., off-centre entrances (Devlin and Naser, 1989). Laymen prefer the reverse (i.e., symmetric order). The architects' preferred options are perceived as more complex, novel, and exciting architecture, while in contrast, laymen prefer the use of more materiality (i.e., a higher degree of materiality in building materials). Architects prefer neutral building materials with less sense of materiality (i.e., building materials that are perceived as more modern and less rustic), centred entrances, and warm colours. These variations are due to the fact that the two groups interpret and appreciate design features differently (Groat, 1982). Beside education, factors influencing aesthetic preference in exterior and interior design features include age, gender, emotional mood, and personality traits (Gifford, 1980). Geographical locations and culture may also play a role here, given such an effect is identified for visual aesthetic preference in digital interfaces (Reinecke and Gajos, 2014). The impact of cultural impact on aesthetic preference finds support in an Asian study by Ghomeshi and Mohd Jusan (2013) that finds no differences in preference between architects and laymen regarding exterior facade features, which is different from the findings of Western studies (Devlin and Naser, 1989; Gifford et al., 2000; Nasar, 1997).

Visual aesthetics – Negative interference or positive stimuli

When the visual environment is overwhelming, visual stress may occur. Visually disturbing elements like clutter and a disorganised setting can disturb us as noise disturbs us auditorily. Hence, it is called visual noise and includes both static and dynamic visual noise (MacKay, 1965). Although visual noise is disturbing, it has been found that interruption effects in cognitive processes, so-called negative re-orientating, is more long-term in acoustic than visual interference (Berti and Schröger, 2001). Visual disturbance is due to its nature, described as a "pop-up" interference, as a visual stimuli that deviates from surrounding, e.g., with regard to colours or orientation (Julesz, 1981). From this perspective, visual interference can be regarded as negative visual aesthetics.

The visual work environment influences employee satisfaction with workplace. Kim and de Dear (2013) found that workplace satisfaction overall was highly correlated with employee visual comfort in different office layouts. The highest level of workplace satisfaction was found among employees in cell-offices (i.e., private office rooms), but there was also higher visual comfort in office layouts with no or low partitions than in those with cubicles with high partitions. The latter result is interesting as it tells us visual comfort is not the same as visual privacy. Kim and de Dear hypothesise that the high rate of positive visual comfort in office layouts with no or low partitions is due to visual connectivity with colleagues. A second hypothesis of Kim and de Dear is that high partitions lead to visual disconnection from surroundings. This prevents employees from determining the source of noise in the office making it uncontrollable or unpredictable for the employees. A complementary hypothesis of mine, based in visual aesthetic perspective, is that the low visual comfort is due to the lack of access to visual environmental stimuli from the surroundings (e.g., view of natural exposure, spacious rooms, colours, and art work). Visual stimulation in school environments has also gained some interest among researchers. A review study of potential health effects of indoor visual environment factors on school children found that of studied factors, art and environmental aesthetics had least impact on the pupils wellbeing (Meng et al., 2023). The researchers found only positive effect on pupil health or academic progress in two studies in the category of indoor visual environment factors (Meng et al., 2023). One of these studies found positive effects on attentional performance and perceived restorative health in biophilicly designed classrooms (Barbiero et al., 2021). While the other found complexity and colour in the decor combined explained 23% of the effect on pupils' academic progress (Barrett et al., 2015).

Visual aesthetics – Personality

The potential association between the individual's personality and preferences in visual aesthetics has attracted some interest in both the field of environmental aesthetics (including visual aesthetics in art and physical environment) and the field of HCI. Investigating visual aesthetics in relation to personality traits, different methods to describe and assess personality are used, e.g., the Eysenck questionnaire and the Big Five personality test (Eysenck and Eysenck, 1975; Goldberg, 1992). The Big Five test uses five broad personality traits – Extraversion/Introversion, Agreeableness, Openness, Conscientiousness, and Neuroticism – to describe personality. It is the method most often used to assess personality by practitioners and researchers. One example of a study using the Big Five is by Furnham and Avison (1997), in which it was found people who score high on Agreeableness have stronger preferences for classical and representational art, and less for abstract art.

Regarding visual aesthetic sensitivity, the research is again contradictory. To exemplify, one study with children (age 10–15 years old) and fine arts students found aesthetic preference to be independent, i.e., not influenced by personality (Frois and Eysenck, 1995), while another found people who scored high on the personality trait of Agreeableness having a higher preference for classical and representational art than for abstract art (Furnham and Avison, 1997). Some diseases may lead to personality change (e.g., Alzheimer's disease (AD)) which may lead to increased egocentrism, mental rigidity, irritation, aggressiveness, etc. How AD influences people's aesthetic preference has therefore gained interest. One study about this found that aesthetic preference differs between participants with AD and those without of the same age (Halpern et al., 2008). Preferences are also stable across time, even though the AD patients had no explicit memory for the paintings tested.

Conclusion

There is a lack of research on visual aesthetics' impact on humans. This is a paradox given that we live in a highly visual society, one that emphasises visual stimuli in contemporary architecture and urban design, that leans on marketing influences design. This is a design approach that leans less on an aesthetics based in craftsmanship or natural design characteristics that are subtly supportive but instead screams loudly "Look at me." Beside the conclusion that there is a lack of empirical research about visual aesthetics and its impact on people, few definite conclusions can be drawn from the presented research on visual aesthetics and its impact on humans with regard to art or physical environment. Nevertheless, based on the research presented, certain conclusions can be drawn. One conclusion is that visual stimulus in comparison to noise has a weaker and more subtle impact. This does not necessarily mean it is not important, but its impact is less clear and more difficult to assess or evaluate. The presented research shows also that various factors like educational background, as well as cultural background and geographical location influence perception of visual aesthetics in different ways. Yet, one important conclusion to be drawn is that humans in general are drawn to orderliness in visual aesthetics. Orderliness can be incorporated in various ways, e.g., in colour scheme or symmetrical order in a physical environment (houses and urban settings) or in work of art it is generally preferred. Why this preference is found is not known, but it appears we need to be trained in design or art to have less preference for this. More knowledge and research about this, as well as other aspects of visual experiences are needed. Perhaps, it is hard to approach the subject, as the matter of beauty is perceived as sensitive or less important. It could also be the fact that there appears to be a difference in taste between those in design professions and laymen, something well studied in terms of architects and non-architects, artists, and non-trained people. This difference seems to apply regardless of geographical location. Although the influence of personality on aesthetic preference has been studied in the different fields of visual aesthetics, no definite conclusions are to be drawn other than personality does not have much impact on our aesthetic preferences.

To summarise, the lack of research on visual aesthetic has the different consequences identified in this chapter, which have various consequences. The scarcity of research does not only result in a lack of an insightful and supportive visual design approach based on empirical evidence, it generates continued fascination with theoretical concepts like Golden Ratio in art and architecture. A fascination that may not be based on facts, but rather is an expression of a desire for something to intrigue us.

Practitioner perspective

Christine Kohlert and Lena Reiß

Introduction

The working world is constantly changing. Digitisation, globalisation, and flexible working models are the driving forces behind this transformation. The result is a growth in employee autonomy and a significant increase in remote working even after the Covid-19 pandemic. However, this change does not only bring advantages. Instead, it is accompanied by rising absenteeism due to mental illnesses such as depression, anxiety, and chronic stress. These diseases also contribute to cardiovascular complaints (Storm, 2022). The significant increase in remote working, whether from home offices or third-party locations such as coworking spaces, presents challenges as well. It is increasingly difficult for companies to maintain team culture and effective communication. Employee wellbeing and productivity decrease as a result (Kohlert, 2021; Microsoft, 2020). Given these challenges, it is crucial for companies to prioritise employee wellbeing. But what does wellbeing encompass, and how can it be effectively promoted in the work environment?

There are various approaches to the definition of wellbeing, whereby subjective wellbeing is decisive. It describes both fleeting and enduring states of happiness (Lambin, 2014). Moreover, wellbeing can be categorised in three dimensions: Physical, social, and mental health (Mayring and Rath, 2013). Social health describes the interaction between individuals in communities and social networks, including the work context. Physical health plays a vital role in determining quality of life, encompassing all aspects of health that affect the physical body. When people are mentally healthy, they can cope with life stressors, use their own abilities, develop themselves further, and play an active role in communities (WHO, 2022). A significant determinant of employee wellbeing is their physical environment (Becker et al., 2022). Thus, a work environment that is tailored to the individual and in which negative factors are minimised has a positive effect on health, wellbeing, and even performance at work (Rohmert and Rutenfranz, 1975). For this reason, it is crucial to consider people's psychological needs during the planning stage and to design space accordingly. This raises the question of what the ideal office should look like to ensure employee wellbeing.

The following sections delve into this matter in detail. Initially, various spatial categories influencing employee wellbeing are introduced. These categories are then clarified and substantiated with practical examples. A comprehensive checklist is then provided, allowing people to evaluate their own office environment.

Categories for the evaluation of wellbeing

To create an office environment in which employees feel comfortable, it is essential to identify the key influencing factors. Employees' requirements for their work environment can be viewed as hierarchical, as illustrated in Figure 4.1.

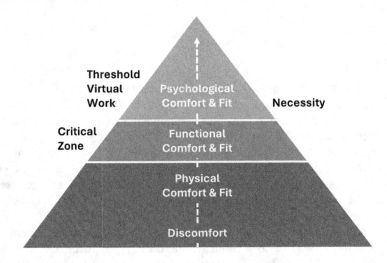

Figure 4.1 Model of satisfaction and wellbeing (Vischer, 2008).

Basic human needs, such as safety and hygiene, must be met so that employees can perform their work. To further enhance the quality of office space, it is crucial to ensure functional amenity. This includes the adaptation of the work environment to the employees' needs. Lastly, it is important to ensure psychological comfort, which entails offering employees a sense of control, belonging, and ownership of their work environment (Vischer, 2008). Furthermore, visual aesthetics influence employee mental state, in combination with other simultaneous sensory experiences (Hekkert and Leder, 2008). Nowadays, physical comfort should be a given in any office environment. Therefore, it makes sense to focus on the upper categories to effectively enhance employee wellbeing.

A set of 16 design parameters, combined to form the wellbeing Score (Reiß, Drees and Sommer), has been defined. These can be measured against a building's features (the object) and their perception by individuals (the subjects) who interact with the space. This comprehensive approach offers insights into how these parameters collectively contribute to promoting wellbeing (Reiß, 2018). The parameters are individuality, innovation, the interior and exterior appearance, privacy, balancing stimulation, community, movement in space, acoustics, light, air, the outdoor space in terms of nature and recreation, sustainability and image, function, orientation, mobility, and safety.

- Individuality covers the presence of flexible furniture as well as lounge areas.
- Innovation entails basing the design on new trends. The interior and exterior appearance should feature zoning and appropriate colours.

- To ensure privacy, the environment should offer retreats and focus zones. Such as small pocket parks, which offer retreats within the space similar to a small cave.
- Balancing stimulation involves variety and contrast in the design of the work environment and the ability to move between crowdy and noisy areas and into peaceful, rather silent zones.
- Community includes the presence of event spaces and a family-friendly design as well as a sense of identity through the presence of groups of people who embody a certain locality. The identity of a place is reinforced by, for example, a certain spoken dialect or customs and events that are celebrated. Collaborative spaces are essential to support this.
- Movement within the space can be encouraged through high-quality circulation and networking spaces. It therefore makes sense to locate a stairwell in a visible and exposed position, to design it to a high standard and, ideally, to equip it with meeting areas, instead of locating the stairwells in a dark corner at the edge of the building, as was the case in the past.
- Acoustics, lighting, and temperature should be pleasant. A range of areas that are designed in different light colours and in which the light colour can be individually adjusted is advantageous. Luminaires equipped with light colours that follow the human biorhythm also contribute to a greater sense of wellbeing. The outdoor space parameter includes providing a view and natural surroundings. Outdoor areas, such as terraces and loggias, should be accessible from a certain distance, e.g., 200 m. So that it is possible to step outside at any time and experience the cold, heat or rain, and wind. This also emphasises the aspect of far-sightedness and views. In combination with the possibility of stepping out, humans are able to confirm themselves in their environment, location, and the seasons at any time.
- Sustainability and image describe the visible corporate image and the sustainability strategy. It makes a difference to wellbeing whether a commitment to sustainability and wellbeing or social aspects can be visibly perceived.
- Sanitary facilities and technical service must be provided to ensure functionality. The design of the sanitary facilities is of particular importance, as these are the rooms that almost everyone, without exception, visits once, if not several times a day. In addition, there is a feeling of utmost privacy, which should be given special attention during the planning process.
- Mobility includes transport links and the provision of parking spaces. It also includes different offers of parking spaces, such as bicycle parking spaces, larger and family parking spaces, and electric car parking spaces.
- The last category, security, involves ensuring security in the office through a security service or an emergency hotline (Reiß, 2018).

Additional design parameters have been established that are complementary and reveal significant overlaps. These include the parameter optimism, which characterises the space's capacity to stimulate creativity and innovation (Kohlert et al., 2018). This can be achieved by granting individuals' control over their work environment and offering opportunities for personalisation, thus overlapping with the design parameters innovation and individuality.

Figure 4.2 Space for mindfulness.

Figure 4.3 Managing cognitive overload.

The second additional design parameter is mindfulness, which describes managing cognitive overload at the workplace by providing retreats and focus zones as well as an environment, the materials, colours, and other features of which are designed to be relaxing. It thus overlaps with the categories of inner and outer appearance and privacy. Individuality, balancing stimulation, and sustainability and image all coincide with authenticity in that they all emphasise the possibility of self-expression in the work environment, the importance of a diverse environment, and the linking of one's own values to those of the company as reflected in the design.

The third category, belonging, focuses on a sense of connection, which can be strengthened by a well-appointed and inviting work environment. Suitable communal spaces are also important. When implementing these parameters, it is crucial not to overlook remote workers. Belonging overlaps with the design parameter "community." Moreover, the work environment should foster a sense of meaning, which entails acknowledging the significance of one's work. This can be cultivated through collaborative and individual workspaces, along with a work environment that supports employees in achieving their goals. These design parameters are also included in individuality and community.

The final additional design parameter, vitality, emphasises the importance of physical health in promoting quality of work. This can be provided by access to nutritious food, the facilitation of physical activity, along with appropriate lighting, temperature, noise level, and a pleasant view. These aspects are also emphasised by the categories of movement in space, acoustics, light, air, and outdoor space. The overlaps are further illustrated in Figure 4.4.

Aside from spatial elements, factors such as leadership, culture, and community also contribute to wellbeing in the office and should not be overlooked. An unfavourable relationship with the manager and the team can amplify an individual's tendency to depression, whereas a good relationship exerts a beneficial influence on overall employee health (Schermuly and Meyer, 2016).

Space can also foster these factors, as demonstrated in the following by a range of practical examples.

Practical examples

In a recent project, we evaluated the wellbeing Scoring system within two distinct workplace environments: a coworking space and a traditional single-office setup. Our study explored the relationship between office design and individuals' perceptions of their adaptability, openness to innovation, creativity, and willingness to embrace change. The coworking space features a variety of functional and aesthetically pleasing offices. Great care was taken to furnish the space with high-quality, visually appealing furniture. Furthermore, the spatial configuration of the coworking space offers a range of open workspaces, retreats and focus zones, communal dining areas, catering facilities, meeting rooms, and a welcoming foyer with concierge service. Collectively, these elements create an atmosphere akin to a residential setting. Conversely, the individual office layout comprises both private and shared offices accommodating two to three workstations.

	Optimum	Mindfulness	Authenticity	Belonging	Meaning	Vitality
Individuality	Individual configurability of the work environment		Scope for self-expression		Other spatial choices	
Innovation	Environment that promotes innovation and creativity					
Inner/outer appearance		Design for tranquillity and freedom from distraction				
Privacy		Provide retreat zones				
Balancing stimulation			Presence of informal variation/ contrast areas			
Community				Strengthen connection to others, inviting atmosphere, common areas	Work towards similar goals in the same place	
Movement in space						Promote exercise, healthy eating options
Acoustics						Pleasant volume
Light						Daylight
Indoor air quality						Pleasant room temperature
Outdoor space/ recreation/nature						Nature experience, pleasant view
Sustainability/ image			Linking own values with company values			Emissions transparency, health and wellbeing, image
Function						Presence of sanitary facilities, technical service, functionality
Orientation						Visibility, recognisability, zoning, hierarchy
Mobility						Accessibility, parking spaces, charging
Safety and security						Security systems/ service, emergency contact
				Include remote workers	Emphasize importance of the work	

Figure 4.4 Convergence of the two sets of design parameters, including the Wellbeing Score (author's representation).

In the Wellbeing Score assessment, the coworking space scored 84 out of 100 points, whereas the individual office layout only scored 43 out of 100. Subsequent user evaluations also confirmed similar ratings across the 16 wellbeing categories. This objectively demonstrates that the diverse coworking environment outperforms the individual office layout by a significant margin, with a doubling of wellbeing levels. These findings are also reflected in users' subjective evaluations. When examining aspects like flexibility, creativity, openness to new ideas, and adaptability to change, it becomes evident that these qualities are closely tied to the functions offered within the office layout. In fact, these four attributes score up to 14% higher in a coworking space compared to a single-office layout. The availability of various work settings – such as spacious discussion tables, focus workspaces with sound insulation, collaborative areas resembling cozy gatherings (potentially on a rooftop terrace), and the opportunity to listen to or participate in conversations, such as those at the coffee machine – is particularly influential (Figure 4.5).

Categories 1	Categories 2	Umbrella Term Consensus	Question	Yes ✓/No ✗
Optimism	Individuality	Individual configurability of the work environment	Can the work environment be adapted to individual requirements?	☐
Optimism	Innovation	Environment that promotes creativity	Does the work environment promote creativity and innovating?	☐
Mindfulness	Inner/outer appearance	Quiet and distraction-free	Can employees find a quiet, distraction-free work environment?	☐
Mindfulness	Privacy	Opportunity to retreat	Does the work environment allow employees to withdraw?	☐
Authenticity	Individuality	Opportunity for self-expression	Does the work environment allow employees to express themselves?	☐
Meaning	Leadership	Meaningfulness of work	Is the significance of employee's work reflected by the work environment?	☐
Authenticity	Balancing stimulation	Areas of variety/ contrast	Are informal areas of variety/ contrast provided?	☐
Belonging	Community	Teamwork	Does the work environment promote teamwork?	☐
Meaning	Community			☐
Authenticity	Sustainability/ image	Positive image	Does the work environment convey company values that are consistent with the employee's own (e.g. sustainability)?	☐
Belonging	Culture, communication	Functioning communication	Does the work environment promote communication (including with remote workers)?	☐
Vitality	Movement in space	Health promotion (physical)	Does the work environment promote physical health (e.g. healthy eating options, promotion of exercise)?	☐
Vitality	Acoustics	Pleasant volume level	Is the noise level in the work environment acceptable?	☐
Vitality	Light	Good lighting	Is the work environment well-lit, with good daylight?	☐
Vitality	Indoor air quality	Pleasant room temperature	Is the temperature in the work environment comfortable?	☐
Vitality	Outdoor space, recreation, nature	Experience of nature	Does the work environment provide aces to nature?	☐
	Orientation	Easy orientation in work environment	Can employees find their way around the work environment (e.g. clarity, zoning, hierarchy)?	☐
	Security	Safety	Is the work environment safe and secure (e.g. security system, emergency contact)?	☐
	Functionality	Functional work environment	Is the work environment functional (e.g. sanitary facilities, technical service)?	☐
	Mobility	Good mobility infrastructure	Does the workplace offer a good mobility infrastructure (e.g. education, parking spaces, charging facilities)?	☐

Figure 4.5 Checklist of office criteria (author's representation).

According to another internal study, these multifunctional and highly varied amenities also have a positive impact on social, mental, and physical wellbeing. Social wellbeing is further increased by the presence of sports, recreational, and creative spaces. Enhanced social wellbeing significantly contributes to an employee's sense of belonging to the company and their level of commitment. As a result, promoting social wellbeing can lead to a remarkable increase of up to 20% in employee commitment and up to 10.5% in their sense of identity with the company. These values, in turn, have a cascading effect on employee turnover rates and overall productivity within the company. The economic implications are of particular interest to company management and can serve as a compelling incentive to implement initiatives aimed at improving the health and wellbeing of their employees.

Conclusion

The examination of practical examples reveals that beyond the favourable impact on people's wellbeing and health, there are also positive effects on economic aspects in a company's work environment. Assessing the positive effects on wellbeing can be accomplished using both quantitative and qualitative methods, such as interviews and surveys. Likewise, the positive influences on economic metrics can be identified by tracking indicators such as illness and turnover rates, as well as engagement levels measured through surveys.

In summary, it can be affirmed that when taking a holistic perspective, adherence to wellbeing principles within the work environment – encompassing both (interior) architectural considerations and an analysis of communication, leadership, and corporate culture – exerts a lasting and beneficial influence on employees and the company's economic performance. These elements are intricately intertwined and of mutual importance.

References

Barbiero, G., Berto, R., Venturella, A. & Maculan, N. (2021). When biophilic design promotes pupil's attentional performance, perceived restorativeness and affiliation with Nature. *Environment, Development and Sustainability*, 1–15.

Barrett, P., Davies, F., Zhang, Y. & Barrett, L. (2015). The impact of classroom design on pupils' learning: final results of a holistic, multi-level analysis. *Building and Environment, 89*, 118–133.

Becker, M., Graf-Szczuka, K. & Wieschrath, S. (2022). Architekturpsychologische Gestaltung von Arbeitsumwelten. *Gruppe Interaktion Organisation Zeitschrift für Angewandte Organisationspsychologie (GIO), 53*(2), 151–159.

Berlyne, D. E. (1970). Novelty, complexity, and hedonic value. *Perception and Psychophysics, 8*, 279–286.

Berti, S. & Schröger, E. (2001). A comparison of auditory and visual distraction effects: behavioral and event-related indices. *Cognitive Brain Research, 10*, 265–273.

Buckley, A. & Jenny, B. (2012). Special issue on aesthetics in mapping — Letter from the guest editors. *Cartographic Perspectives, 73*(67), 3–11.

Carlson, A. (2000). *Aesthetics and the Environment: The Appreciation of Nature. Art and Architecture*. London, New York: Routledge.

Cela-Conde, C. J., Marty, G., Munar, E., Nadal, M. & Burges, L. (2002). The 'style scheme' grounds perception of paintings. *Perceptual and Motor Skills, 95*, 91–100.

Da Luz Reis, A. T. & Dias Lay, M. C. (2010). Internal and external aesthetics of housing estates. *Environment and Behavior, 42*(2), 271–294.

Devlin, K. & Naser, J. (1989). The beauty and the beast: some preliminary comparisons of "high" versus "popular" residential architecture and public versus architect judgements of same. *Journal of Environmental Psychology, 9*, 333–344.

Dewey, J. (1980). *Art as Experience*, New Edition. New York: Perigee books.

Ebenhard, J. P. (2009). *Brain Landscape: The Coexistence of Neuroscience and Architecture*. New York, USA: Oxford University Press.

Eysenck, H. J. & Eysenck, S. B. G. (1975). *Manual of the Eysenck Personality Questionnaire*. London: Hodder & Stoughton.

Frois, J. P. & Eysenck, H. J. (1995). The visual aesthetic sensitivity test applied to Portuguese children and fine arts students. *Creativity Research Journal, 8*(3), 277–284.

Furnham, A. & Avison, M. (1997). Personality and preference for surreal paintings. *Personality and Individual Differences, 23*(6), 923–935.

Furnham, A. & Walker, J. (2001). The influence of personality traits, previous experience of art, and demographic variables on artistic preference. *Personality and Individual Differences, 31*, 997–1017.

Ghomeshi, M. & Mohd Jusan, M. (2013). Investigating different aesthetic preferences between architects and non-architects in residential façade designs. *Indoor and Built Environment, 22*(6), 952–964.

Gifford, R. (1980). Judgments of the built environment as a function of individual differences and context. *Journal of Man-Environment Relations, 1*(1), 22–31.

Gifford, R., Hine, D., Muller-Clemm, W., Reynolds, J. R. & Shaw, K. T. (2000). Decoding modern architecture: a lens model approach for understanding the aesthetic differences of architects and laypersons. *Environment and Behavior, 32*(2), 163–187.

Goldberg, L. R. (1992). The development of markers for the Big-Five factor structure. *Psychological Assessment, 4*(1), 26–42.

Groat, L. (1982). Meaning in post-modern architecture: an examination using the multiple sorting task. *Journal of Environmental Psychology, 2*(1), 3–22.

Halpern, A. R., Ly, J., Elkin-Frankston, S. & O'Connor, M. (2008). 'I know what I like': stability of aesthetic preference in Alzheimer's patients. *Brain and Cognition, 66*, 65–72.

Hekkert, P. & Leder, H. (2008). Product aesthetics. In H. Schifferstein & P. Hekkert (Eds.), *Product Experience*. New York: Elsevier, pp. 259–285.

Hekkert, P. & Wieringen, P. C. W. (1996). Beauty in the eye of the expert and nonexpert beholders: a study in the appraisal of art. *American Journal of Psychology, 109*, 389–407.

Jacobsen, T., Schubotz, R. I., Höfel, L. & von Cramon, D. Y. (2006). Brain correlates of aesthetic judgment of beauty. *Neuroimage, 29*, 276–285.

Julesz, B. (1981). Textons, the elements of texture perceptions, and their interactions. *Nature, 290*, 91–97.

Kaplan, S. (1992). *Environmental Preference in a Knowledge-Seeking, Knowledge-Using Organism*. Oxford: Oxford University Press.

Karsvall, A. (2002). Personality preferences in graphical interface design. *Paper Presented at the ACM International Conference Proceeding Series*, 217–218. New York: ACM.

Kawabata, H. & Zeki, S. (2004). Neural correlates of beauty. *Journal of Neurophysiology, 91*(91), 1699–1705.

Kim, J. & de Dear, R. (2013). Workspace satisfaction: the privacy-communication trade-off in open-plan offices. *Journal of Environmental Psychology, 36*, 18–26.

Kohlert, C. (2021). *Das menschliche Büro - The Human(e) Office: Hilfe zur Selbsthilfe für eine gesunde Arbeitswelt - Helping People to a Healthy Working Environment*. Wiesbaden: Springer Fachmedien Wiesbaden.

Kohlert, C., Cooper, S., & Drees & Sommer (2018). *Space for Creative Thinking: Design Principles for Work and Learning Environments*. Munich: Callwey GmbH.

Lambin, E. (2014). *Glücksökologie. Warum wir die Natur brauchen, um glücklich zu sein*. Hamburg. French original edition titled Une Écologie du Bonheur. Paris: Éditions Le Pommier.

Lang, J. (1987). *Creating Architectural Theory: The Role of the Behavioral Sciences in Environmental Design*. New York, NY: Van Nostrand Reinhold.

MacKay, D. M. (1965). Visual noise as a tool of research. *The Journal of General Psychology, 72*(2), 181–197.

Martindale, C., Moore, K. & West, A. (1988). Relationship of preference judgments to typicality, novelty, and mere exposure. *Empirical Studies of the Arts, 6*, 79–96.

Mayring, P. & Rath, N. (2013). *Glück – Aber Worin Liegt Es? Zu Einer Kritischen Theorie des Wohlbefindens*. Göttingen: Vandenhoeck & Ruprecht.

Meng, X., Zhang, M. & Wang, M. (2023). Effects of school indoor visual environment on children's health outcomes: a systematic review. *Health & Place, 83*(103021), 1–18.

Microsoft (2020). *Building Resilience & Maintaining Innovation in a Hybrid World: Top-five Practical Steps Leaders Can Take Today.* https://grekai.files.wordpress.com/2021/06/wp-1624687760426.pdf

Moore, G. T. & Golledge, R. G. (Eds.) (1976). *Environmental Knowing: Theories, Research and Methods.* Stroudsburg, PA: Dowden, Hutchinson & Ross.

Nadal, M., Capo, E. M. A., Rosello, J. & Cela-Conden, C. J. (2008). Towards a framework for the study of the neural correlates of aesthetic preference. *Spatial Vision, 21*(3–5), 379–396.

Nasar, J. (1992). Connotative meanings of house styles. In E. Arias (Ed.), *The Meaning and Use of Housing.* London: Gower.

Nasar, J. L. (1997). New developments in aesthetics for urban design. In G. T. Moore & R. W. Marans (Eds.), *Toward the Integration of Theory, Methods, Research, and Utilization. Advances in Environment, Behavior and Design,* vol. 4. Boston: Springer.

Naser, J. L. (Ed.) (1988). *Environmental Aesthetics: Theory, Research, and Applications.* Cambridge: Cambridge University Press.

Neperud, R. W. (1986). The relationship of art training and sex differences to aesthetic valuing. *Visual Arts Research, 12,* 11–19.

O'Connor, Z. (2006). Bridging the gap: Façade colour, aesthetic response and planning policy. *Journal of Urban Design, 11*(3), 335–345.

Pelowski, M., Markey, P. S., Forster, M., Gerger, G. & Leder, H. (2017). Move me, astonish me… delight my eyes and brain: the Vienna integrated model of top-down and bottom-up processes in art perception (VIMAP) and corresponding affective, evaluative, and neurophysiological correlates. *Physics of Life Reviews, 21,* 80–125.

Reiß, L. (2018). *Der Transitraum am Hub Flughafen: Analyse und Auswirkungen der Raumqualität und des Raumcharakters des Umsteigeweges auf das Wohlbefinden des/der Umsteigepassagiers/in.* Darmstadt: Technische Universität Darmstadt.

Reinecke, K. & Gajos, K. Z. (2014). Quantifying visual preferences around the world. *Paper Presented at the CHI 2014, 32nd Annual ACM Conference on Human Factors in Computer Systems.* Toronto, Canada.

Rohmert, W. & Rutenfranz, J. (1975). *Arbeitswissenschaftliche Beurteilung der Belastung und Beanspruchung an unterschiedlichen industriellen Arbeitsplätzen.* BMA Referat Öffentlichkeitsarbeit. Bonn: Bundesminister für Arbeit und Sozialordnung.

Salimun, C., Seman, E. A. A., Ahmad, W. N. W., Abdullah Sani, Z. H. & Shishehchi, S. (2021). Predicting aesthetic preferences: does the big-five matters? (IJACSA) *International Journal of Advanced Computer Science and Applications, 12*(12), 175–182.

Sartwell, C. (2017). Beauty. Metaphysics Research Lab. *The Stanford Encyclopedia of Philosophy.* Archived from the original on 26 February 2022. Retrieved 26 May 2021.

Schermuly, C. C. & Meyer, B. (2016). Good relationships at work: the effects of leader-member exchange and team-member exchange on psychological empowerment, emotional exhaustion, and depression: good relationships at work. *Journal of Organizational Behaviour, 37*(5), 673–691. https://doi.org/10.1002/job.2060.

Silvia, P. J. (2005). Cognitive appraisals and interest in visual art: exploring an appraisal theory of aesthetic emotions. *Empirical Studies of the Arts, 23,* 119–133.

Song, J., Kwak, Y. & Kim, C.-Y. (2021). Familiarity and novelty in aesthetic preference: the effects of the properties of the artwork and the beholder. *Frontiers in Psychology, 12,* 1–17.

Stone, N. J. & English, A. J. (1998). Task type, posters, and workspace colour on mood, satisfaction, and performance. *Journal of Environmental Psychology, 18*(2), 175–185.

Storm, A. (2022). Beiträge zur Gesundheitsökonomie und Versorgungsforschung: Gesundheitsreport 2022. Heidelberg: medhochzwei Verlag GmbH.

Ulrich, R. S. (1983). Human behavior and environment: advances in theory and research. In I. Altman & J. F. Wohlwill (Eds.), *Behavior and the Natural Environment*, vol. 6. New York: Springer, pp. 85–125.

Vischer, J. (2008). Towards an environmental psychology of workspace: how people are affected by environments of work. *Architectural Science Review, 51*(2), 97–108.

World Health Organization (2022). *World Mental Health Report: Transforming Mental Health for All*. World Health Organization.

Yina, J., Zhua, S., MacNaughtona, P., Allena, J. G. & Spenglera, J. F. (2018). Physiological and cognitive performance of exposure to biophilic indoor environment. *Building and Environment, 132*, 255–262.

Zhong, W., Schröder, T. & Bekkering, J. (2022). Biophilic design in architecture and its contributions to health, wellbeing, and sustainability: a critical review. *Frontier in Architectural Research, 11*, 114–141.

5   Lighting

Editors' introduction

Purpose-built workplaces created today for knowledge workers are lit with electric lighting. Many workplaces use as much natural light as they can throughout the day. However, deep floor plans, windowless rooms or blinds-down behaviour means that light switches are often flipped on to supplement natural daylight.

Electric lighting has many parameters, such as how bright it is and what colour it is, which can complicate planning and maintaining lighting schemes for workplaces. However, we can partly simplify the planning by bringing it back to its most basic and thinking about our early days as a species when the light bulb was unimaginable. Then, light was dimmer and warmer and lower on the horizon (or literally on the ground in a fire) at the beginning and end of the day and brighter and cooler at noon when the sun was overhead. As mentioned in Chapter 2, we continue to find dawn/dusk-type lighting a relaxing experience; it can also enhance our ability to get along with others and to think creatively. Similarly, when light is brighter and cooler, our ability to concentrate on the task-at-hand, be efficient, etc., gets a positive jolt. The circadian lighting movement is based on these fundamental responses to the light in our world.

DOI: 10.1201/9781003390848-5

Researcher perspective

Jennifer Veitch and Naomi Miller

Introduction

The advent of electric lighting in the late 19th century together with other advances in building engineering changed the shape of buildings and the way we use them. Buildings became taller, and their footprints became wider, increasing occupant distances from windows; and, in some places, windows became smaller and non-opening as mechanical ventilation came into use. Workdays extended the darker mornings and evenings as reliance on daylight became less. In the early 20th century, professional associations for lighting came into being in individual countries and internationally. These bodies continue today to meld light and lighting science with lighting practice to develop recommendations and standards that guide the development of lighting products, their application and regulations to protect the public and to provide a level playing field for industry.

In the beginning, electric light provided the means to see the work, whether it was sums to be added or buttons to be sewn. Over time, it became clearer that lighting for architectural interiors has many goals, and these may be met with both electric lighting and day lighting, often together. Here are a few:

- providing light to support the work we do,
- wayfinding within the building,
- providing safety from tripping and falling,
- enhancing visibility where personal security might be a concern,
- illuminating the faces of people to support communication,
- providing accent on signage and artwork,
- making spaces appear cheerful and inviting,
- highlighting architectural form and beautiful materials,
- revealing the true colour of objects, surfaces and people
- and, ideally, affecting mood by delivering drama, delight and distraction.

Various jurisdictions recognise voluntary, and in some cases mandatory, standards for workplace lighting to achieve these goals (Comité Européen de Normalisation (CEN), 2021; Commission Internationale de l'Eclairage (CIE), 2001; Illuminating Engineering Society (IES), 2020). All of them implicitly or explicitly seek to deliver lighting that meets a definition of lighting quality that can be summarised graphically in a "three circles" model (Figure 5.1). Good lighting must meet the needs of its users, which are more than mere detail detection, but must do so while respecting architectural form and composition together with economic, practical and environmental considerations including energy use, maintenance and cost. Integrating these requirements into a single installation makes the lighting designer's job challenging.

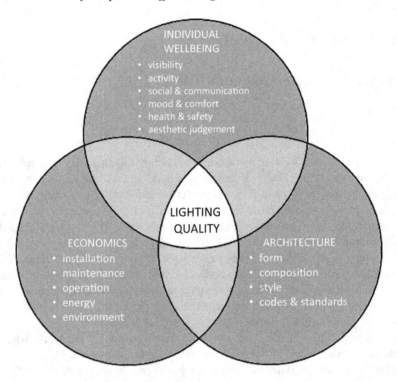

INDIVIDUAL
WELLBEING
• visibility
• activity
• social & communication
• mood & comfort
• health & safety
• aesthetic judgement

LIGHTING
QUALITY

ECONOMICS
• installation
• maintenance
• operation
• energy
• environment

ARCHITECTURE
• form
• composition
• style
• codes & standards

Figure 5.1 The "three circles" model of lighting quality (Veitch et al., 1998) underlies the internationally-accepted definition of lighting quality.

A brief detour into the science of light detection

Perceptive readers will have noticed that all of the goals listed above are met through vision. At the light levels typical of interiors, the most important photoreceptors (the cells in the retina that detect light) are the cones, which are concentrated in the fovea near the centre of the retina. The three cone photoreceptors have peak photodetection in the short-, medium- and long-wavelength range of the visual spectrum (roughly, 448 nm [blue-appearing], 541 nm [green-appearing] and 569 nm [red-appearing], respectively) and are packed close together. The cones provide both colour vision and the detection of fine details. Another photoreceptor cell, the rod, dominates the rest of the retina; rod cells are active at low light levels with peak activation at 507 nm. There isn't space here to provide details about vision, but one good source about how this is important for lighting is Boyce (2014).

Around the end of the 20th century, our understanding of photoreception was revolutionised with the discovery of another class of photodetector, the *intrinsically photoreceptive retinal ganglion cell* (ipRGC) (Berson et al., 2002). These cells connect the eye to many brain structures outside of the visual cortex, including the suprachiasmatic nucleus of the hypothalamus, where the body's principal internal clock resides.

The first action spectrum to describe how the light detected from the ipRGCs influences the night-time release of an important hormone, melatonin, was published in 2001 (Brainard et al., 2001; Thapan et al., 2001). The nightly rise and fall of melatonin synchronise many physiological functions including digestion, immune function and the sleep-wake cycle. With refinements, the CIE published a standard to describe the spectral sensitivity of the ipRGCs, establishing their peak sensitivity at 490 nm (CIE, 2018) – clearly, not the same as the classical photoreceptors. This means that the established way to measure the quantity of light, photopic illuminance, is not the correct quantity to describe ipRGC light exposures. This is because photopic illuminance is based on the $V(\lambda)$ spectral luminous efficiency function (which itself is a combination of the medium- and long-wavelength-cone sensitivity). This realisation led to the definition of new quantities to describe light exposures. Of particular relevance here, the CIE system of metrology for ipRGC-influenced effects of light-established quantities for five photoreceptor types in relation to the equivalent exposure to the CIE daylight illuminant D65 (an agreed-upon light spectrum that mimics clear-sky noon daylight), known as the *α-opic equivalent daylight illuminance (EDI)* reported in the unit lux (lx). "α" refers generically to any of the five photoreceptor types: three cones, the rod or the ipRGC. For the ipRGCs, the relevant quantity is melanopic EDI.

ipRGCs are integrators of light exposure, detecting the amount of light in the environment. There is more to be learned, however, as various laboratories have shown that there are several subtypes of ipRGC with different spectral sensitivity functions (Guido et al., 2022). Furthermore, the ipRGCs, cones, and rods work together to influence vision and a host of other physiological and psychological processes; for example, ipRGCs influence pupil size (Güler et al., 2008). This is a very active research area, with hundreds if not thousands of new research papers published annually, so expect recommendations to change as we learn more about how light affects more of our biology than only what we see.

Back to goals for lighting

With the additional knowledge about the role that light plays as a regulator of human physiology and psychology, there are new goals for lighting designers to consider. Some general truths are clear even today:

- a period of bright light exposure to the eye every day is necessary for good health; and,
- a period in true darkness every night is equally necessary.

To date, there is one recommendation for a specific pattern of daily light and darkness (Brown et al., 2022). Using the CIE system to describe the measurement quantities for ipRGC-influenced effects of light (CIE, 2018), this group of international photobiology experts concluded that one ought to be exposed to 250 lx of melanopic equivalent daylight illuminance (melanopic EDI) throughout the day; 10 lx melanopic EDI for three hours before bedtime; and, not more than 1 lx melanopic EDI at night while sleeping. These light levels are not measured on the desktop or the floor but rather measured vertically at the eye in the direction in which one looks.

The fact that healthful light exposure relies on a daily pattern means both that workplace lighting should contribute to that pattern, and that individuals also have some responsibility for their own light hygiene.

What about sustainability?

Electric lighting is a substantial contributor to building energy use, even with the widespread adoption of light-emitting diode (LED) lighting. Most jurisdictions place limits on the allowable level of energy used for electric lighting (American Society of Heating, Refrigerating, and Air Conditioning Engineers (ASHRAE), 2022; Canadian Commission on Building and Fire Codes (CCBFC), 2020). To achieve the target light exposures suggested by Brown et al. (2022) with electric lighting only while meeting these energy limits is very difficult, if not impossible (Safranek et al., 2020). In those parts of the world where daylight availability is mandated, achieving a healthful light exposure by blending daylight and electric light will be easier, but as we shall see, successful integration requires skilful design. In places where daylight design is optional, and at times of year when daylight is in naturally short supply, using general lighting to meet the light exposure targets might not be practical.

What does all this mean for workplace lighting in the 21st century?

Integrative lighting is the phrase used for "lighting integrating both visual and non-visual effects, and producing physiological and/or psychological benefits upon humans" (CIE, 2020). Good workplace lighting should strive, so far as possible, to deliver this integration. Our focus here is on lighting for daytime workplaces.

Light levels. Minimum guidance for light levels (illuminance on a working plane) is provided in various expert recommendations, standards and sometimes workplace regulations. The ISO/CIE 8995-*1* standard for indoor workplaces (currently under revision, publication expected in mid- 2024) and the European EN 12464-*1* standard (CEN, 2021) both provide target values for a wide variety of workplace types. The Illuminating Engineering Society publishes an online Illuminance Selector tool that finds target illuminance values from any of its application standards (which are published as separate documents). In all of these documents, the illuminance values are based on making tasks visible, and not on other health or behaviour outcomes.

Exceeding these minimum values, and seeking also to provide the target daytime melanopic EDI, while respecting local energy regulations, will require an expert designer and careful choices. For example, the target melanopic EDI need not be provided in spaces that are not occupied for long periods (e.g., corridors, washrooms). Automated controls can ensure that electric light is provided only when spaces are occupied, and only to the extent that is needed in addition to available daylight. Working with the interior design and architectural team to ensure that the larger surfaces have light colours will help to deliver the light to the occupants' eyes. Using white or very light colours on most surfaces, using more saturated tones only for visual interest and emphasis, is preferable. Light surfaces also contribute to a feeling of spaciousness, which is generally preferred (Bokharaei and Nasar, 2016; Stamps, 2010).

Light sources. Light should contain a healthy broad spectrum throughout the day. This will achieve several goals. Not only will it deliver energy in the wavelengths to which ipRGCs are sensitive (violet, blue and cyan-appearing light), but it will also include energy across the spectrum so that the colour rendering of objects is excellent. Including red light will keep people looking healthy. High colour rendering also contributes to better visual performance (Papamichael et al., 2016). Look for a light source with a colour rendering index (CIE R_a) of 90 or higher with R_9 values ≥ 75 for best results.

Ideally, the light source should also include some near-infrared radiation (NIR). There is some evidence that NIR can contribute to healing and mitochondrial function, and one recent experiment found that adding some NIR exposure to the skin improved mood and daytime drowsiness (Giménez et al., 2023).

Note that there is no single "best" light source spectrum, so long as a white-appearing light is used. The lighting designer can use their best judgement on this point, taking into account the overall design intent. Different regions have local preferences; in Asia, it can be more common to see a high correlated colour temperature (5000 K or higher), whereas in Europe and North America, a lower (warmer) light source colour is more typical (3500 or 4000 K).

One of the best light sources to achieve all of these goals is daylight. Daylight is an intense, broad-spectrum light source that includes NIR, although some wavelengths are attenuated by window glass. Along with these benefits, using daylight also provides critical cues on time of day, and weather. Daylight varies over the day in colour appearance (spectrum), solar angle and intensity. The variability itself is a source of interest.

Windows can also provide views and connect people to activities in the outside world while offering the opportunity for distant views that can rest the eyes. The connection to the outside world, particularly if the view is natural and/or attractive, brings psychological benefits (Galasiu and Veitch, 2006).

With the advent of LED lighting and the diversity of ways in which it can be controlled, contemporary electric light sources have the potential to cause problems related to temporal light modulation (TLM), which is the general phrase describing variations in their light output over time. TLM can sometimes be seen as flicker but can also cause a moving object to appear jerky (stroboscopic effect) or can cause a light source or a lighted object to appear as a repeating image as the eyes scan across it (phantom array effect). Some TLM can cause headaches, eyestrain, disrupted reading and changed brain activity (Wilkins et al., 2010). In Europe, regulations that limit TLM in all LED light sources at full power come into effect in late 2024 (European Commission, 2021, February 26), but there are no similar regulations in many other parts of the world, including North America.

Light fixtures (luminaires). There is good evidence that, overall, people prefer lighting systems that combine direct (down) light and indirect (up) light (Boyce et al., 2006a; Veitch and Newsham, 1998). However, there are large individual differences in preferences for light level (Newsham and Veitch, 2001) and it seems that these differences influence the preference for the light direction. People who prefer brighter rooms value the mix of direct and indirect light, whereas those who prefer a lower light level are indifferent to differences in light direction (de Vries et al., 2021). The amount of light on the walls is also important; alertness is higher when the walls are more brightly lit (de Vries et al., 2018).

Glare. Glare is to light as noise is to sound. Too much light in the wrong place can cause discomfort and can make details less visible. The visibility loss occurs both because of light scattered in the eye (think of seeing an obstacle on the road when oncoming headlights shine in your eyes), and because of light diffused across the surface of a shiny object (as when reading a glossy magazine in direct sunlight). Discomfort arises when there is a very high contrast or a very high light intensity in the field of view. Preventing both adverse outcomes is a matter of careful selection of light fixtures, where they direct their light, and their location. Interestingly, the discomfort is greater when the excessive light comes from electric lighting than when daylight is the source (Hopkinson, 1970). Other contextual and individual differences, including task engagement and the presence of an interesting view outside, also influence the experience of discomfort, despite decades of effort by lighting researchers to predict discomfort solely on the basis of physical parameters (Kent et al., 2017).

Lighting controls. Lighting controls can be very simple – a switch on a wall – or very complex, with automated presence detection, dimming in response to the available daylight, and even the potential to change the light source colour appearance (a feature known variously as "dynamic lighting" or "colour-tenable lighting"). Some lighting controls are mandated as part of energy codes. Unquestionably, automated controls save energy and reduce peak electrical demand (Galasiu et al., 2007). What is perhaps less well appreciated is that individual control over workplace lighting, at the level of the occupant's workstation, adds to the energy savings (Galasiu et al., 2007) and has other benefits for individuals and organisations, described below.

Practitioner perspective

Naomi Miller and Jennifer Veitch

How do lighting designers use all this?

The lighting designer's task is a balancing act, in which competing needs and requirements trade off against one another and sometimes against other architectural considerations. The starting point is always the function and the characteristics of the people who will occupy the space (Figure 5.2 and Table 5.1).

Table 5.1 Questions and best practice

Questions and best practices	Why this is important
Function	
What are the functions of the space? Is it an office? A sports facility? A mine? A house of worship?	The answers to these questions provide a first approximation to defining the needed intensity and three-dimensional patterns of the light distribution.
What does the user need to see, and where?	
Visual tasks and their visual-spatial locations may include reading numbers on a vertical computer screen, seeing the dirty floor while mopping, seeing codes and retrieving materials from vertical racks in a warehouse, assembling small parts onto a PC board on a horizontal counter, or chopping onions in the kitchen. They can include reading peoples' faces and expressions or administering medications to healthcare patients.	
Occupants	
Who will be using the space? What are their ages and visual capabilities? Do they have any special needs or neurological sensitivities, such as a sensitivity to flicker or a need for subdued light (e.g., albinism)? Do they have aging eyes needing more light and time transitioning between dim indoor and bright outdoor conditions?	These questions refine the targets for the necessary light intensity, and also contribute to the lighting choices for delivering appropriate light for visual comfort, safety, and for supporting daily biological health.

(*Continued*)

Table 5.1 (Continued)

Questions and best practices	Why this is important
Daylight	
Is it a new building? What is the potential to influence the availability of daylight? For renovations of existing buildings, can existing window openings be enlarged? Can skylights be added?	The lighting designer's aim is to bring in a high level of daylight for as many hours as possible. New buildings have more freedom to take advantage of daylight apertures because windows and skylights can be optimised for interior uses and engineered to maximise building envelope efficiency and minimise leaking and other maintenance issues.
What is the potential to influence the choice of window materials? Best practice is to use clear, highly efficient glazing materials that maximise light transmission over the window area (VLT of 60% or greater). Be aware that coatings and tints may attenuate an important part of the spectrum, so consider triple-pane glass with no low-e coatings as a compromise between efficiency and broad spectral transmission, for example. Skylights may benefit from a slight diffusion on the glass to soften the direct beams of sunlight.	The lighting designer aims to make the most possible use of daylight. This can be a negotiation with other engineers and architects because of the effect that the window choices have on overall building energy performance.
Given the building, its location, orientation, and climate, what are the best ways to control glare from daylight? Effective means of glare control include overhangs and interior light shelves, strategic planting of trees, diffusion on skylights or splayed openings that reduce the contrast between the skylight and the surrounding ceiling.	Effective façade and roof design minimises glare and dramatically reduces electric light use. Reducing the need for blinds and curtains contributes to more use of daylight because glare prompts the occupants to draw the interior blinds, but they seldom remember to open them after the bothersome sun angle has changed. There is increasing awareness of the need to control the amount of light escaping from buildings at night to protect other animals, birds, and insects whose physiology needs darkness as much as humans do.
Light levels	
Check local regulations for the space type and occupant characteristics to choose light level targets, and factor in any energy regulations.	The target levels guide subsequent choices of light fixtures (luminaires), their placement (layout) and controls.

(Continued)

Table 5.1 (Continued)

Questions and best practices	Why this is important
In general, this will mean, for most office or classroom spaces, a target of 500 lx of maintained photopic illuminance at the workplane, and if possible 250 lx melanopic EDI at the eye. Task lighting can be used to supplement any overhead lighting systems to achieve the desktop target where occupants work. Achieving the suggested target of 250 lx melanopic EDI vertically at the eye will usually require daylight to supplement electric light. Light levels in corridors can be much lower (in the neighbourhood of 100 lx). Illuminating artwork or walls rather than the floor contributes to increasing the vertical illuminance and visual comfort. In circulation spaces, contrast of materials used in signage, stairs, ramps, and doorways is more important than light levels.	

Interior finishes

What are the plans for interior finishes on walls and ceilings? Best practice is to use mostly white or light colours for major surfaces, and to use deep colours sparingly as accents. Similarly, wood finishes absorb light and can reduce the amount of light delivered to the viewer.	White or light-coloured (high reflectance) surfaces bounce light many times, diffusing light throughout the space and supporting energy efficiency while providing a bright, cheerful appearance. The aperture of a light fixture (luminaire) blends better with its surround when the ceiling is white, and this is more comfortable for viewers.

Light fixtures (luminaires)

Given the activities and viewer needs, what are the right choices for light fixtures in the various occupied areas? For general lighting, aim to use luminaires or combinations of luminaires that deliver a mixture of indirect (up) light and direct (down) light. Recessed or pendant downlights can create harsh shadows. If downlights are used, supplement with lighting that reflects off room surfaces, which helps diffuse the shadows. Wall sconces, wall washers or indirect lighting provide excellent supplementary lighting.	Choosing the correct combination and locations of luminaires delivers the required quantity of light for the activities in the space and also the required vertical light at the occupant's eye, making it look attractive. At the same time, luminaire placement helps to avoid conditions that cause discomfort. A high proportion of uplight increases the light levels vertically at the occupant's eye, contributing to meeting the target melanopic EDI, and does so without uncomfortable excess light from bright lenses or apertures.

(*Continued*)

Table 5.1 (Continued)

Questions and best practices	Why this is important
Directly visible light sources in workplaces should direct the light primarily straight down to the work surface (e.g., desk or countertop). This downlight should be designed such that the downward distribution is concentrated from 0 to 45° from the luminaire nadir (See Figure 5.2). In the sketch, the cut-off angles show the optical cut-offs one would use to prevent glare from reaching the viewer's eyes, where it could cause discomfort.	The direct light on the task makes it more visible. Limiting the direction and avoiding any view of exposed light sources prevents discomfort.
Consider adding task lights for focused light on desks or workbenches in specific areas. Task lights efficiently supplement light levels where difficult visual tasks are performed.	Task lights deliver light when and where it is needed, and are one way to adjust to individual needs.
Avoid luminaires with exposed light sources, or luminaires with very unevenly-bright apertures. Strong patterns such as stripes of LEDs, or linear luminaires with high-brightness narrow apertures can be painful to view.	

Light sources

What light source will best reveal the colours that are expected to be in the space?	Creating a pleasant appearance benefits occupant mood and wellbeing.
Are there cultural considerations in the specification of light source colour appearance (correlated colour temperature)?	Meeting local expectations contributes to occupant acceptance.
Is there budget to increase the colour rendering of the light source from the minimum ($R_a \sim 80$) to the preferred ($R_a > 90$)?	Higher colour rendering sources improve visual performance and enhance colour discrimination, but they can come at a higher purchase price.

Lighting controls

What lighting controls shall be used?	Light levels in most spaces will need to be adjusted in output based on daylight availability, visual preferences, and task requirements. Presence detection ensures that the lights consume energy only when occupants are present.
At a minimum, all luminaires should be dimmable. Some jurisdictions also mandate presence detectors. Smart lighting systems integrated with building energy management systems add daylight dimming and other functions.	Turning off interior lights in unoccupied spaces at night also prevents ecological damage from unnecessary light exiting from windows.

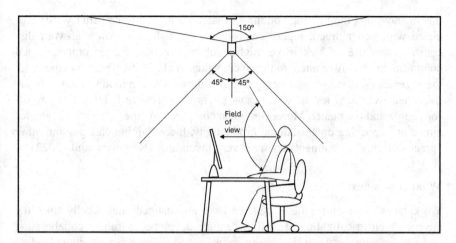

Figure 5.2 Sketch of a seated worker at a desk under a suspended direct/indirect luminaire (image by Naomi Miller).

Isn't this complicated? Why should we care about workplace lighting?

Humans are visual creatures. Most people obtain the majority of their information about the world from what they see. Clearly, workplace lighting is necessary to perform the work, whether that work is typing on a keyboard, fastening bolts on an airframe or examining a patient's vital signs. These visual requirements underpin all lighting recommendations and standards, which mostly express the requirements as specific target light levels on the working surface.

Workplace lighting that meets these minimum requirements has excluded truly bad lighting from most places (Boyce, 2014). We can, however, do better – and when we do, the result is better for workers, their organisations and the environment. Better lighting uses direct/indirect luminaires, to provide a higher vertical illuminance as well as lighting task surfaces, and it incorporates individual control over local light levels. These individual controls can be unpopular with building managers because they are more complex to install and to maintain, but they pay off. There is a wide range of individual differences in preferred light levels (Boyce et al., 2006b; Veitch and Newsham, 2000), which means that any single target light level, at best, will satisfy 50% of the population (Newsham and Veitch, 2001). Working under one's preferred light level (even if one did not personally choose the setting) is associated with better mood and higher ratings of lighting quality (Newsham and Veitch, 2001), and the only way to ensure that individuals have this opportunity is to give them individual control.

Laboratory studies have shown that working under lighting conditions that one judges to be of higher quality benefits mood, environmental satisfaction, at-work visual discomfort and work motivation (Veitch et al., 2008; Veitch et al., 2013). A field investigation over several months in a functioning organisation extended this to show that the lighting conditions that were judged to be better were associated with better mood, higher environmental satisfaction, lower at-work discomfort, less use of sick leave, higher job satisfaction, higher organisational commitment and lower intent to turnover (Veitch et al., 2010). Taken together with the observation that lighting energy use also drops when individuals have choice over their own light levels, we see that paying attention to lighting can pay off for people and the planet. Moreover, when compared to other corporate strategies aimed at improving organisational productivity, these lighting changes and other "green building" investments compare very favourably (Newsham et al., 2022).

What comes next?

Workplaces, work and lighting systems have all changed dramatically since the advent of electric lighting. The first century developed a robust collaboration and conversation between lighting researchers and lighting practitioners, leading to a strong evidence base supporting the design process of lighting primarily for vision. The second century will bring increasing depth of understanding about the physiological and psychological effects of light exposure, particularly through the ipRGCs. Expect the collaboration and conversation between researchers and practitioners to bring further elaboration of lighting and day lighting equipment, light sources, design tools and public policy to support more fully integrative lighting that genuinely fits the definition of high-quality lighting.

References

American Society of Heating, Refrigerating, and Air-Conditioning Engineers (ASHRAE) (2022). *Energy standard for sites and buildings except low-rise residential buildings* (ANSI/ASHRAE/IES Standard 90.1-2022). Atlanta, GA: ASHRAE.

Berson, D. M., Dunn, F. A. & Takao, M. (2002). Phototransduction by retinal ganglion cells that set the circadian clock. *Science, 295*(5557), 1070–1073.

Bokharaei, S. & Nasar, J. L. (2016, November 1). Perceived spaciousness and preference in sequential experience. *Human Factors, 58*(7), 1069–1081.

Boyce, P. R. (2014). *Human factors in lighting* (3rd ed.). Abingdon: CRC Press.

Boyce, P. R., Veitch, J. A., Newsham, G. R., Jones, C. C., Heerwagen, J. H., Myer, M. & Hunter, C. M. (2006a). Lighting quality and office work: Two field simulation experiments. *Lighting Research and Technology, 38*(3), 191–223.

Boyce, P. R., Veitch, J. A., Newsham, G. R., Jones, C. C., Heerwagen, J. H., Myer, M. & Hunter, C. M. (2006b). Occupant use of switching and dimming controls in offices. *Lighting Research and Technology, 38*(4), 358–378.

Brainard, G. C., Hanifin, J. P., Greeson, J. M., Byrne, B., Glickman, G., Gerner, E. & Rollag, M. D. (2001). Action spectrum for melatonin regulation in humans: Evidence for a novel circadian photoreceptor. *Journal of Neuroscience, 21*(16), 6405–6412.

Brown, T. M., Brainard, G. C., Cajochen, C., Czeisler, C. A., Hanifin, J. P., Lockley, S. W., Lucas, R. J., Münch, M., O'Hagan, J. B., Peirson, S. N., Price, L. L. A., Roenneberg, T., Schlangen, L. J. M., Skene, D. J., Spitschan, M., Vetter, C., Zee, P. C. & Wright, K. P., Jr. (2022). Recommendations for daytime, evening, and nighttime indoor light exposure to best support physiology, sleep, and wakefulness in healthy adults. *PLoS Biology, 20*(3), e3001571.

Canadian Commission on Building and Fire Codes (CCBFC) (2020). National energy code of Canada for buildings. National Research Council of Canada.

Comité Européen de Normalisation (CEN) (2021). *Light and lighting - Lighting of work places – Part 1: Indoor work places* (EN 12464-1). Paris, France: CEN.

Commission Internationale de l'Eclairage (CIE) (2018). *CIE system for metrology of optical radiation for ipRGC-influenced responses to light* (CIE S026/E:2018). Vienna, Austria: CIE.

Commission Internationale de l'Eclairage (CIE) (2020). ILV: International lighting vocabulary, 2nd Edition (CIE S 017/E:2020). Vienna, Austria: CIE. https://cie.co.at/e-ilv.

de Vries, A., Heynderickx, I. E. J., Souman, J. L. & de Kort, Y. A. W. (2021). Putting the ceiling center stage – The impact of direct/indirect lighting on room appraisal [Article]. *Building and Environment, 201*.

de Vries, A., Souman, J. L., de Ruyter, B., Heynderickx, I. E. J. & de Kort, Y. A. W. (2018). Lighting up the office: The effect of wall luminance on room appraisal, office workers' performance, and subjective alertness. *Building and Environment, 142*, 534–543.

European Commission (2021, February 26) Commission Regulation (EU) 2019/2020 of 1 October 2019 laying down ecodesign requirements for light sources and separate control gears pursuant to Directive 2009/125/EC of the European Parliament and of the Council and repealing Commission Regulations (EC) No 244/2009, (EC) No 245/2009 and (EU) No 1194/2012. *Official Journal of the European Union, 68*, L68/108–148.

Galasiu, A. D., Newsham, G. R., Suvagau, C. & Sander, D. M. (2007). Energy saving lighting control systems for open-plan offices: A field study. *LEUKOS, 4*(1), 7–29.

Galasiu, A. D. & Veitch, J. A. (2006). Occupant preferences and satisfaction with the luminous environment and control systems in daylit offices: A literature review. *Energy and Buildings, 38*(7), 728–742.

Giménez, M. C., Luxwolda, M., Van Stipriaan, E. G., Bollen, P. P., Hoekman, R. L., Koopmans, M. A., Arany, P. R., Krames, M. R., Berends, A. C., Hut, R. A. & Gordijn, M. C. M. (2023). Effects of near-infrared light on wellbeing and health in human subjects with mild sleep-related complaints: A double-blind, randomized, placebo-controlled study. *Biology, 12*(1), 60.

Guido, M. E., Marchese, N. A., Rios, M. N., Morera, L. P., Diaz, N. M., Garbarino-Pico, E. & Contin, M. A. (2022). Non-visual opsins and novel photo-detectors in the vertebrate inner retina mediate light responses within the blue spectrum region. *Cellular and Molecular Neurobiology, 42*(1), 59–83.

Güler, A. D., Ecker, J. L., Lall, G. S., Haq, S., Altimus, C. M., Liao, H.-W., Barnard, A. R., Cahill, H., Badea, T. C., Zhao, H., Hankins, M. W., Berson, D. M., Lucas, R. J., Yau, K.-W. & Hattar, S. (2008). Melanopsin cells are the principal conduits for rod-cone input to non-image-forming vision. *Nature, 453*(7191), 102–105.

Hopkinson, R. G. (1970). Glare from windows - 2 what people say. *Construction Research and Development Journal, 2*(4), 169–175.

Illuminating Engineering Society (IES) (2020). American national standard practice for office lighting (ANSI/IES-RP-1-2020). New York: IES.

Kent, M. G., Altomonte, S., Wilson, R. & Tregenza, P. R. (2017, February 15). Temporal effects on glare response from daylight. *Building and Environment, 113*, 49–64.

Newsham, G. R. & Veitch, J. A. (2001). Lighting quality recommendations for VDT offices: A new method of derivation. *Lighting Research and Technology, 33*, 97–116.

Newsham, G. R., Veitch, J. A., Zhang, M. Q. & Galasiu, A. D. (2022). Comparing better building design and operation to other corporate strategies for improving organizational productivity: A review and synthesis. *Intelligent Buildings International, 14*(1), 3–22.

Papamichael, K., Siminovitch, M., Veitch, J. A. & Whitehead, L. (2016). High color rendering can enable better vision without requiring more power. *LEUKOS, 12*(1–2), 27–38.

Safranek, S., Collier, J. M., Wilkerson, A. & Davis, R. G. (2020). Energy impact of human health and wellness lighting recommendations for office and classroom applications. *Energy and Buildings, 226*, 110365.

Stamps, A. E., III. (2010). Effects of permeability on perceived enclosure and spaciousness. *Environment and Behavior, 42*(6), 864–886.

Thapan, K., Arendt, J. & Skene, D. J. (2001). An action spectrum for melatonin suppression: Evidence for a novel non-rod, non-cone photoreceptor system in humans. *Journal of Physiology, 535*(Pt 1), 261–267.

Veitch, J. A., Julian, W. & Slater, A. I. (1998). A framework for understanding and promoting lighting quality. In J. A. Veitch (Ed.), *Proceedings of the First CIE Symposium on Lighting Quality* (Vol. CIE-x015-1998, pp. 237–241). Commission Internationale de l'Eclairage.

Veitch, J. A. & Newsham, G. R. (1998). Lighting quality and energy-efficiency effects on task performance, mood, health, satisfaction and comfort. *Journal of the Illuminating Engineering Society, 27*(1), 107–129.

Veitch, J. A. & Newsham, G. R. (2000). Preferred luminous conditions in open-plan offices: Research and practice recommendations. *Lighting Research and Technology, 32*, 199–212.

Veitch, J. A., Newsham, G. R., Boyce, P. R. & Jones, C. C. (2008). Lighting appraisal, wellbeing, and performance in open-plan offices: A linked mechanisms approach. *Lighting Research and Technology, 40*(2), 133–151.

Veitch, J. A., Newsham, G. R., Mancini, S. & Arsenault, C. D. (2010). *Lighting and office renovation effects on employee and organizational wellbeing* (NRC-IRC RR-306). Ottawa, ON: NRC Institute for Research in Construction.

Veitch, J. A., Stokkermans, M. G. M. & Newsham, G. R. (2013). Linking lighting appraisals to work behaviors. *Environment and Behavior, 45*(2), 198–214.

Wilkins, A. J., Veitch, J. A. & Lehman, B. (2010). LED lighting flicker and potential health concerns: IEEE Standard PAR1789 update. In *Proceedings of the Energy Conversion Congress and Exposition (ECCE) 2010 IEEE, 12–16 Sept., 2010* (pp. 171–178). Institute of Electrical and Electronics Engineers. https://doi.org/10.1109/ECCE.2010.5618050.

6 Acoustics, noise, and soundscapes

Editors' introduction

Noise (unwanted sound) is a common occurrence in the office workplace with the sounds of background intelligible speech, from people interacting and meeting nearby, or mechanical sounds, such as from printers/copiers and coffee machines, being two key sources of distraction. Historically, noise in the office was mostly due to overheard telephone conversations, but today the distracting speech comes from people interacting and collaborating in-person or online at their desk.

The fact that noise distraction can be due to interaction makes it challenging to create workplaces that reduce the undesirable effects of noise on performance and workplace satisfaction, because effective collaboration is key to any organisation's success. In addition to adding sound absorption in the workplace (using ceiling tiles, wall panels, and furniture, etc.), creating zones for noisier activities away from ones that require quieter conditions is another useful solution. However, success depends on colleague behaviours and how reliably they take noisy activities (like meetings, telephone, and online calls) out of earshot of their colleagues or in areas with effective acoustic shielding.

In many workplaces, a key method used to prevent noise from degrading well-being and performance involves partitions or fully enclosed spaces. Not necessarily private offices, but meeting rooms and focus pods or booths etc. However, when people "camp out" in small rooms provided for short bursts of intense activity, their colleagues can become resentful or stressed (van Meel, 2019). So, in addition to physical solutions, behaviour is fundamental to reducing noise in offices.

Reducing noise distraction in offices, especially dense open-plan ones, is one of the biggest challenges for the workplace design and management community.

DOI: 10.1201/9781003390848-6

Researcher's perspective

Nick Perham

Irrelevant sound effect

Laboratory research on auditory distraction has been a prominent area of cognitive research that reveals how sound impacts performance and provides greater insight into the cognitive processes involved therein. Two sources of auditory distraction have been identified with their overarching nomenclature termed the duplex account (Hughes, 2014). One source, the irrelevant sound effect (ISE) derives from interference to humans' linguistic abilities to produce and articulate language, whereas the other source, the deviance effect, stems from our general ability to be susceptible to changes in our auditory environment and its power to capture our attention.

Arguably the most researched auditory distraction phenomenon in the laboratory is the ISE. First observed in the late 1970s (Colle and Welsh, 1976), it shows the depreciable drop, around 30%–50%, in short-term memory performance in the presence of a background sound condition compared to a quiet, control condition (see Jones, 1999, for a general overview). Interestingly, the ISE occurs whether the sound is played during presentation and recall of the items or just during presentation (Miles et al., 1991), is independent of the intensity of the sound so performance is similar whether the sound is as loud as a shout or as quiet as a whisper (Tremblay and Jones, 1999), occurs whether participants like the sound or not (Perham and Vizard, 2010), does not require participants to understand the content of the sound if it is speech (Jones et al., 1990), and, moreover, the sound does not even need to be speech (Jones and Macken, 1993).

Typically, the task used to test distraction involves serial recall, also known as digit span, in which a series of seven to nine to-be-recalled (TBR) items is presented at the rate of about one per second and recalled in strict serial recall position, that is, the order in which the items were presented. Two features are keys to the ISE. First, the emphasis on recalling in serial recall position promotes the use of participants' rehearsal abilities that help to maintain and retrieve the order information necessary to successfully complete the task. However, this act of rehearsal is impaired by the background sound resulting in the ISE. This is the second key feature. Specifically, sounds that have been shown to elicit the ISE are those that contain acoustical variation, termed changing-state, where each successive sound item is temporally discrete from the next, such as speech (up to three or so voices, also known as the "babble" effect; Jones and Macken, 1995), most music (apart from music where the sounds blur into a homogenous mass such as "grind-core metal" from bands like Repulsion and Napalm Death, Perham and Sykora, 2012), and many background sounds in our auditory environment such as vehicles, machinery, electronic devices, people moving around, doors opening and closing, and even animals.

The differences between these items are processed without our awareness to determine whether they derive from the same spatial location (Bregman, 1990) and it is the order information from these differences which conflict with the order information inherent in rehearsing the of the TBR items that impairs performance. In contrast, sounds that contain little or no acoustical variation, termed steady-state, such as lots of speech (three or more voices at the same time; Jones and Macken, 1995), repeated sound items (Jones et al., 1995), white or pink noise (Ellermeier and Zimmer, 1997; Smith et al., 1981), and blurred music (Perham and Sykora, 2012), elicit little or no disruption. Thus, for the changing-state effect to be observed, a conflict of processing order information must occur such that the task requires rehearsal (e.g. free recall, mental arithmetic, relying on memory to identify a particular option, language learning, relying on memory to copy a visual display; Beaman and Jones, 1998; Perham and Macpherson, 2012; Perham et al., 2007; Perham et al., 2013; Perham et al., 2016; Saffran et al., 1999; Waldron et al., 2011) and the sound contains changing-state information. When either of these properties is absent such as in the missing-item task (identifying which one of an exhaustive set, such as days of the week, are missing; Jones and Macken, 1993) or when the sound is more steady-state in nature, then the changing-state effect is smaller than typically observed or not observed at all.

A similar conflict of process has also been noted with semantic, rather than, order information. Here, the processing of semantic information, such as in reading a passage of text and answering multiple choice questions about the text or outputting presented items by categories rather than serial order, is impaired more when the content of the sound is processed semantically than when it is not (Marsh et al., 2008, 2009; Martin et al., 1988; Perham and Currie, 2014). In contrast to the physical properties of the sound, such as phonology, linguistic stimuli are processed to the deeper level of their meaning. Debate surrounds whether these stimuli, like the physical properties, can be processed without attention although some recent evidence of greater recognition for ignored, deviant items (see deviant effect below) suggests that attention to the ignored items may take place (Perham et al., 2023).

Although the most successful explanation of the ISE focuses on this conflict in processing order information, the ISE also seems to be affected to some degree by participants' attention to the sound or the task. That is, certain speech, such as one's own name (Röer et al., 2013), more emotional words (Buchner et al., 2004), or taboo words (Röer et al., 2017), produce a larger changing-state effect suggesting that when the sound/speech is more salient/controversial it produces additional impairment. However, the finding that attentional capture is still observed when the task does not require rehearsal, for example in the missing-item task (Hughes et al., 2007), indicates that attentional cannot explain the classic changing-state effect inherent in the ISE and that these "semantic" effects are examples of a general attentional diversion mechanism that is not specific to serial processing.

Specific and aspecific deviant sound

While the majority of the second source of auditory distraction has generally been researched more recently than the ISE, it has a longer evolutionary history. It concerns our general hearing capabilities which predate our linguistic skills and is, arguably, shared with many other hearing animals. More specifically, it occurs when one sound item differs from the others and, in doing so, captures our attention. In some situations, this may be a deliberate attempt to focus on an individual sound, such as an individual speaker in a room full of speakers, as in Cherry's "cocktail party" effect (Cherry, 1953), or, for non-human animals like frogs, insects, and some birds, an individual signaller amid the myriad of other similar acoustic signals (Bee and Micheyl, 2008). Other times, the individual may be focusing on the focal task and ignoring the sound, but this deviant sound may break through the auditory status quo accidentally, or deliberately in the case of an auditory alarm such as in medical situations (Edworthy, 2013), and capture the individual's attention.

This latter, breakthrough situation has more recently been the focus of auditory distraction researchers and describes what is known as the deviance effect (Hughes, 2014). Evidence for its existence and impact is observed, similarly to the ISE, in a depreciable drop in serial recall performance. However, this time the explanation arises from the momentary, at least, diversion of mental resources away from the primary task to the deviant sound due to attentional capture.

Two forms of this effect have been identified – specific and aspecific. In both forms, attention is captured by the deviant sound item but in the former, it derives from its meaningfulness to the individual (Röer et al., 2013), such as their name or something that they desire (e.g. food) or are concerned about (e.g. spiders), whereas for the latter, it is context rather than the content of the deviant that drives its power (Hughes et al., 2007), for example, the single letter B amongst a series of As - there is nothing about its content that is salient, but its attentional power derives from its ability to violate the expectation of the series being all As.

Evidence for the distinction between the two sources of auditory distraction arrives from multiple findings. For example, aspecific deviant sounds impair per-formance for tasks that do not require rehearsal, such as the missing-item task, whereas changing-state sounds do not (Hughes et al., 2007). Further, specific devi-ant sounds, such as valenced/emotional or taboo words, exert their effects on both seriation and non-seriation-based tasks in contrast to the changing-state effect that only reveals itself on rehearsal-based tasks (Marsh et al., 2018; Rettie et al., 2023).

On an individual level, individuals with a higher working memory capacity, as measured by reading (Daneman and Carpenter, 1980) and operation spans (Turner and Engle, 1989), are shown to be less vulnerable to distraction (Conway and Kane, 2001; Kane and Engle, 2003). This greater inhibitory ability reveals them to be less affected by deviant sounds than those individuals with lower working memory capacity (Hughes et al., 2013; Sörqvist, 2010), but equally susceptible to the changing-state effect (Elliott and Briganti, 2012; Hughes et al., 2013; Neath et al., 2003; Sörqvist, 2010). Finally, there is developmental evidence showing that children are more affected by attentional capture than changing-state (Elliott et al., 2016). When children (approximately aged seven years old) and adults were given both a probe-order task (similar to serial recall except that at output the participant was presented with the name of a TBR item and asked to name the TBR that directly followed it) and a missing-item task, both groups displayed the changing-state effect in the probe-order task but not for the missing-item task. What was more revealing was that children's missing-item task performance was greatly reduced by both steady- and changing-state sounds whereas adults' performance was not, suggesting that children may be vulnerable to any sound rather than sounds with a particular quality such as changing-state.

Content of background sound

More recent developments have focused on the potential influence of the content of the background sound. That is, despite clear instructions to ignore the background sound while performing the focal task, speculation has arisen regarding as to whether the content of the sound is actually processed and, if so, how much influence it may have on later behaviour. Although this influence has been revealed by Cherry (1953), its validity has been questioned due to the both the task and sound being presented in the same modality. The ISE resolves this issue by involving a visually-presented task in the presence of sound, that is, stimuli from two modalities. Evidence suggests that the content of the background sound can influence behaviour both implicitly (spelling) and explicitly (recognition). When asked, as part of a seemingly unrelated task, to spell words following an ISE experiment, participants responded with the non-dominant spelling of homophones that formed the background sound more than their dominant-spelling counterparts (Richardson et al., 2023). Also, in a deviance experiment where all the background sounds were words, participants recognised the background deviant words three times as often as non-deviant words presented in the same auditory position (7th of 9) resulting in a decrease in recall performance of the 7th and the following 8th TBR item (Perham et al., 2023). Further, recognition of unpresented items was greater if they were semantically related to the deviant than if they were not.

Conclusion

In summary, laboratory research into auditory distraction reveals the underpinning mechanisms and processes of the everyday experience of navigating an often-changing complex auditory environment in which concentration and communication are vulnerable to auditory sources that compete with, and distract us from, our goals. The implications of such research are far-reaching in terms of its population and application.

Practitioner's perspective

Paige Hodsman

Introduction

The literature relating to acoustics in offices dates back decades (Beranek, 1956; Broadbent, 1955). Subsequent research regarding performance, health, wellbeing, and overall satisfaction is extensive (Andringa and Lanser, 2013; Banbury and Berry, 2005; Nemecek and Grandjean, 1973). However, putting the research into practice remains elusive for most workplace design professionals, and acoustic satisfaction remains low, even in the highest-performing offices (Leesman, 2023).

Acoustics

Now, more than ever, organisations rely on the optimisation of cognitive processing (Fisher et al., 2017). Offices must be designed to support work activities, including communication and individual focus. Although seemingly opposed, these two cognitive processes are fundamentally linked to auditory processing. Thus, the acoustic environment becomes a vital enabler for optimal performance (James, 2021; Hughes and Jones, 2003; Weber et al., 2022). These conflicting activities explain, in part, why open-plan offices are notoriously challenging to treat acoustically to any significant degree of occupant satisfaction (James, 2021; Harvie-Clark and Hinton, 2021; Oseland and Hodsman, 2018).

Human beings react to sound differently. Auditory functioning can be affected by genetics, environment, injuries, ageing, and temporary conditions like a cold or ear infection. Additionally, offices often have multi-lingual speakers, where the acoustic environment can significantly affect the ability to hear speech clearly and focus on concentration tasks. Finally, ever-increasing awareness and knowledge regarding personality traits, auditory sensitivities, and neurodiversity in the workplace warrant a more serious prioritisation of the acoustic design of offices (Pichora-Fuller et al., 2016; van der Heijden et al., 2019).

Recent studies have identified three primary areas of concern for office workers relative to acoustic satisfaction: Control, privacy, and disruptions from other people talking. Several key factors also affect noise perception: Task and work activity, perceived control and predictability, context and attitude, personality type, and sensory sensitivity (Harvie-Clark and Hinton 2021; James et al., 2021; Oseland and Hodsman, 2020).

Noise

Noise is an unwanted sound, and the noises perceived in offices are often those one lacks autonomy over and are a distraction from the task at hand. This is particularly problematic when individual focus work is taking place. Hughes and Jones (2003) state that we use processing power to interpret the sounds around us. Even irrelevant information, although ignored, is still actively being processed, thus reducing the ability to focus on the work task and ultimately affecting performance (Hughes and Jones, 2003); the ISE is discussed in the researcher section of this chapter.

Controlling sound for communication and individual focus across the open office space, through walls and ceiling voids, is a central feature of a well-designed office. Human beings are particularly good at hearing in the speech frequencies. It is difficult to "tune out" speech, making the auditory environment of offices challenging to control (Banbury and Berry, 1998).

Control is vitally important for providing a sense of security, reduced anxiety and essential for one's wellbeing (Leotti et al., 2010). Increased satisfaction with the acoustic conditions can be attributed to individual control over the work environment. This aspect was investigated in a study by Harvie-Clark and Hinton (2021), where auditory satisfaction, environmental comfort, and the feeling of control were all closely linked to productivity in the workplace.

Measuring sound

Room acoustics encompasses sound transmission (how sound travels from room to room, through the air and materials), insulation (blocking of sound from room to room, into the open plan or from outside), absorption (essentially the assimilation of sound energy as opposed to reflecting), sound reflection (sound waves bouncing off of objects), and sound diffusion (sound waves scattering in multiple directions after hitting an object). Room acoustics also includes how humans perceive sounds in different settings.

One of the primary challenges in office acoustics is determining what to measure, which can vary depending on the room's occupants, activities, and physical characteristics. Different types of spaces require different approaches; for instance, a small meeting room may focus on sound insulation and reverberation time for clarity, while a large open-plan office, with its added complexity of size, shape, materials, and multiple activities, will require an approach to minimise speech propagation, overall sound pressure level (SPL), and reverberation time (RT).

SPL is one of the most common measurements in room acoustics, and research has shown it relates to stress levels. A cross-over design field experiment by Seddigh et al. (2015) showed lower cognitive stress in better acoustic office conditions versus worse conditions with higher background sounds. This study found that small reductions in SPL showed a subjective, self-reported effect on cognitive stress. Another study showed that higher sound levels contributed to physiological stress, specifically both steady-state and impulse sound exposure, which elevated stress responses compared to quiet sounds (Radun et al., 2022). High SPLs also contribute to the Lombard Effect (Brumm and Zollinger, 2011), whereby voice sound volumes will increase as the background sound levels increase, resulting in very high SPLs in the room. While important, however, SPL is not the only factor to consider.

RT refers to the time it takes for the sound to decrease by 60 decibels (dB) after the sound source stops over a specified period, e.g. 20 seconds.

Like SPL, RT alone can neither adequately describe the acoustic condition, nor predict the outcome for occupant comfort levels in open-plan offices (Nilsson, 2004; Nilsson et al., 2008). Despite this, many standards and guidelines focus on SPL and RT.

In office acoustics, C_{50} refers to one of the parameters used to measure the sound decay in a room. It quantifies the time it takes for the sound level to reduce by 50 decibels (dB) after the sound source has stopped; this is desirable where clear communication is essential. C_{50}, in conjunction with the speech transmission index (STI), can optimise the acoustic performance of the space.

STI is a measure used to evaluate speech intelligibility. It quantifies listeners' speech understanding by considering background noise, reverberation, and other acoustic disturbances. The impact of various factors on speech clarity can be assessed by measuring the STI.

The attenuation of speech $(D_{A,S})$ refers to the reduction or decrease in the loudness or intensity of someone's speech. It can occur due to distance from the listener, background noise, or physical elements, such as furniture, limiting sound transmission. Attenuation affects the clarity and intelligibility of speech, making it harder for the listener to hear and understand, which is needed to reduce the spread of unwanted speech.

The term "spatial decay rate of speech" $(D_{2,S})$ refers to the rate at which sound intensity decreases as it moves away from the source. The decay rate of speech can be affected by the presence of obstacles, reflections, and the acoustic characteristics of the surrounding space. This is important for reducing the spread of speech.

So far, the measurement discussion has been centred around open-plan offices, as most tend to be open-plan. However, controlling sound within smaller rooms and insulating the sound from transferring from room to room is equally important, if not more so, for creating privacy (Bradley and Gover, 2010).

Insulation from internal airborne noise, or Difference in Normalised Transmission, A-weighted (adjusted for human hearing), (DnT,A), refers to the sound insulation rating of a partition or structure. Essentially, DnT,A provides information about how well a partition/wall can attenuate sound across various frequencies, allowing for the assessment of its soundproofing capabilities.

Additionally, normalised flanking level difference (Dn,f) is a standardised rating that measures the sound insulation performance of building components like doors, windows, walls, or floors. It assesses how well soundproofing materials or structures reduce sound transmission between different areas.

Despite best efforts to provide standards and guidance for office acoustics, the measurements often only tell part of the story. A persistent criticism pertains to acoustic measurements being taken without the presence of people, considering perceptions (other than A-weighting) and behaviours. Other considerations are needed to create acoustically comfortable spaces that support the occupants' work.

ISO 22955:2021 is a new comprehensive standard that attempts to more fully address the complexities of open-plan office acoustic performance requirements and offers guidelines for designing office spaces regarding noise control and speech intelligibility. It is based on six types of spaces defined by activities and recommends using surveys to gain evidence of the occupants' perceptions beforehand.

It is important to note that, apart from this, there are no universal guidelines, standards, or regulations for office acoustics. Different countries will have different criteria, so referring to the specific locale's standards/regulations before applying any acoustic solution is best.

Soundscape

Just as most objects are "acoustic" (meaning they have acoustic properties), all offices will have a soundscape, but few soundscapes are deliberate by design. In simple terms, a soundscape is a collection of sounds that can be heard in a particular environment (Brown et al., 2016). The International Standard for Soundscaping (International Organization for Standardization, 2014) aims to enable a broad international consensus on the definition of "soundscape" to provide a foundation for communication across disciplines and professions.

The soundscape may include conversations, keyboard strokes, phone calls, printer sounds, background music, and heating ventilation and air conditioning (HVAC) equipment. Offices are often noisy, yet equally, they can be extremely quiet. Deliberate soundscape solutions will incorporate a range of sounds, including low sound levels, based on the occupants' needs. A deliberately designed soundscape will use an evidence-based approach that considers the people, activities taking place, and the physical properties of the building structure, materials and existing interiors. Soundscapes can be assessed using the grounded theory approach (Acun and Yilmazer, 2018). Data is gathered through interviews, observations, and other methods to derive theories based on evidence rather than preconceptions. Acoustic mapping can aid soundscape design by highlighting areas where auditory activity is most intense (see Figure 6.1). "Liveliness" mapping based on activities shows promise as a basis for soundscape design (Harvie-Clark and Hinton, 2021; Vellenga et al., 2017). Oseland and Hodsman (2018) developed an activity-based acoustic design methodology that uses four layers based on the occupants, the activities, and the physical room and uses qualitative and quantitative data input to formulate the basis for practical solutions (see Figure 6.2).

As highlighted previously, various factors influence the perception of sound, and defining "unwanted sound" is subjective. Reactions to noise extend beyond loudness and are influenced by psychological and personal factors. To ensure high satisfaction levels, it becomes crucial to consider occupants' perceptions, especially when introducing masking sounds deliberately into the office soundscape. Sound masking, broadly speaking, can use a continuous, ambient background sound of mixed frequencies or natural sounds like flowing water to mask or cover up distracting sounds. This can be an effective way to control distractions (Haapakangas et al., 2011). Hongisto (2008) found that a significant improvement in objective speech privacy occurred after installing a sound masking system. The radius of distraction (how far sound can travel before it interrupts someone's concentration) was reduced from 13 to 6 meters. Although noise disturbance declined, the change was not statistically significant.

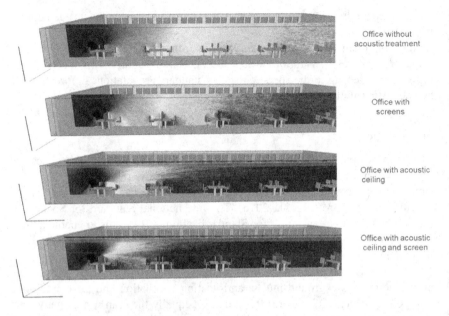

Office without
acoustic treatment

Office with
screens

Office with acoustic
ceiling

Office with acoustic
ceiling and screen

Figure 6.1 "Liveliness" mapping based on activities.

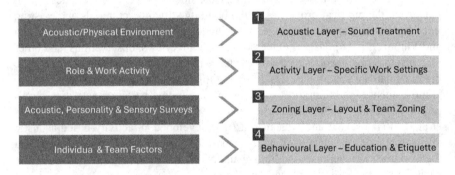

Figure 6.2 Acoustic zoning and layers.

However, only some respond positively to sound masking, and for some individuals, it can result in annoyance (Lenne et al., 2020). Brown et al. (2016) tell us there is a difference between creating acoustic environments in which sounds are introduced regardless of individuals' sensitivities and soundscapes. The introduction of additional sound can affect how people perceive the acoustic environment of the space. When designing soundscapes, personal sensitivity, context, activities, and the existing physical environment must be considered (Lenne et al., 2020).

Several studies have explored the impact of nature sounds (Haapakangas et al., 2011; Luo et al., 2021; Ratcliffe et al., 2013). Gould van Praag et al. (2017) used functional brain imaging (fMRI) to compare naturalistic and artificial sound responses; their findings suggest that health benefits could be derived from exposure to natural environments. Other studies have demonstrated mixed results when used in office settings (Hongisto et al., 2017; Jahncke et al., 2016). Whether people will respond positively to introducing additional sound in the soundscape is subjective. For example, the sound of flowing water is generally perceived as favourable by people, but this could psychologically be construed negatively depending on individual experiences or when it is introduced without the proper context, for example, interpreting water sounds as "leaking pipes or toilets" (Hongisto et al., 2017). Favourable responses to bird sounds may be dependent on the type of bird and cultural meanings (Ratcliffe et al., 2013).

A study by Hongisto et al. (2017) explored the effects of water-based sound masking systems versus traditional pseudo-random masking in an open-plan office. The results suggested that water-based sound masking was perceived less well. However, the authors concluded that water-based sounds could not be excluded for masking sounds in offices.

A study by Abdalrahman and Galbrun (2017) examined specific sources of water-based sound masking to mask irrelevant speech and found that preferences for water sounds depended on the type of water sound and the presence of a visual stimulus. Some water sounds were beneficial when accompanied by a visual water stimulus, while others were not.

Advancements in understanding neural processing are shedding light on sound source location capabilities. This may help aid the creation of more comfortable acoustic spaces. Research on the auditory receptors in the inner ear and further processing in the brain indicate that humans have spatial awareness and expectation contextually for where sounds should come from (Raghunand et al., 2016; van der Heijden et al., 2019). For example, if using artificial sounds of a babbling brook or stream, placing the sound source near the floor to mimic where the sound would be expected in nature and including a visual representation is advisable.

Whether artificial or natural sounds are introduced into the soundscape, due to the unique capabilities of human hearing, it is important to use high-quality, well-researched technology sensitive to the human auditory system. For example, avoid repeated sounds or "looping," make the sounds random. When using natural sounds, choose those that mimic the local biosphere and are reflective of the time of day and temperature (Benfield et al., 2014; Raghunand et al., 2016; van der Heijden et al., 2019). Also, it is essential to provide control through on/off switches, the ability to select the type of space and sounds, and volume control.

The inconclusive nature of sound masking studies further emphasises the need to assess the space's occupants before embarking on the acoustic design solution. This is particularly important considering individual differences, including neurodiverse conditions and auditory sensitivity (Font-Alaminos et al., 2020).

Implementation

- *Increase awareness and knowledge* – The evidence-based approach can improve knowledge and general understanding of the importance of good acoustic conditions and be used to help persuade decision-makers to allocate adequate resources.
- *Impressions of cost* – Good acoustic solutions can be perceived as expensive. If the client and broader project team understand the benefits of a good acoustic solution, securing a satisfactory outcome becomes exceedingly more accessible. There is evidence that supports the potential financial impact that the acoustic design can have on the organisation. Oseland et al. (2022) developed a return on workplace investment tool to calculate the impact of interior environmental conditions and determined the return on a high-level investment in acoustic solutions to be 6.7%, the highest of any other factor.
- *Which products to use* – There are standards for testing acoustic materials, for example, ASTM C423 or ISO 354:2003 *Acoustics – Measurement of Sound Absorption in a Reverberation Room* (International Organization for Standardization, 2003). However, there needs to be more monitoring on how or if the data is reported accurately in product data released to the public. Most objects and materials in an office will have inherent acoustic properties. Labelling a product as "Acoustic" tells little to nothing about how it performs. This ambiguity can be confusing and, in some instances, misleading to those who specify materials. Therefore, more explicit, transparent information is needed to help specifiers determine the best products. Consistent terminology and transparent data sheets combined with the knowledge of interpreting them are all necessary to determine which products will perform best. The recent advent of the supply of environmental product declarations in certified building schemes such as LEED (2019), BREEAM (2018), and WELL Certified (2018) may add to the much-needed practice of transparency.
- *Checks and balances to learn and improve* – Using pre- and post-occupancy questionnaires dedicated to the acoustic condition is vital for feeding the research base and for continual improvement. However, some reservations about using profiling tools based on cultural norms or data protection legislation could exist. The ISO 22955:2021 (International Organisation for Standardisation, 2021) recommends the use of questionnaires and references the GABO, French acronym for Acoustic Annoyance in Open-Plan Offices (Pierrette et al., 2014) acoustic survey to determine occupant satisfaction with the physical working environment, noise assessment and sensitivities, and general health.

The complicated interplay of people, activities, building physics, and organisational objectives set a complex criterion for creating optimal office acoustic solutions. But with a human-centred, evidence-based approach, offices of the future will support, rather than hinder, the efforts of the people in the space so that they can do the jobs they were hired to do.

References

Abdalrahman, Z. & Galbrun, L. (2017). Audio visual preferences of water features used in open plan offices. *In Proceedings of the 24th International Congress on Sound and Vibration,* July 23–27, London.

Acun, V., & Yilmazer, S. (2018). A grounded theory approach to investigate the perceived soundscape of open-plan offices. *Applied Acoustics, 131,* 28–37.

Andringa, T. C. & Lanser, J. J. L. (2013). How pleasant sounds promote and annoying sounds impede health: A cognitive approach. *International Journal of Environmental Research and Public Health, 10*(4), 1439–1461.

Banbury, S. & Berry, D. C. (1998). Disruption of office-related tasks by speech and office noise. *British Journal of Psychology, 89*(3), 499–517.

Banbury, S. P. & Berry, D. C. (2005). Office noise and employee concentration: Identifying causes of disruption and potential improvements. *Ergonomics, 48*(1), 25–37.

Beaman, C. P. & Jones, D. M. (1998). Irrelevant sound disrupts order information in free as in serial recall. *Quarterly Journal of Experimental Psychology, 51A,* 615–636.

Bee, M. A. & Micheyl, C. (2008). The "cocktail party problem": What is it? How can it be solved? And why should animal behaviourists study it? *Journal of Comparative Psychology, 122*(3), 235–251.

Beranek, L. L. (1956). Criteria for office quieting based on questionnaire rating studies. *The Journal of the Acoustical Society of America, 28*(5), 833–852.

Benfield, J. A., Taff, B. D., Newman, P., Smyth, J. M. & Moga, M. M. (2014). Natural sound facilitates mood recovery. *Ecopsychology, 6*(3), 183–188.

Bradley, J. S. & Gover, B. N. (2010). Speech levels in meeting rooms and the probability of speech privacy problems. *The Journal of the Acoustical Society of America, 127*(2), 815–822.

BREEAM (2018). *Building Research Establishment Environmental Assessment Method.* Retrieved from: https://www.breeam.com/

Bregman, A. S. (1990). *Auditory Scene Analysis: The Perceptual Organisation of Sound.* Cambridge: MIT Press.

Broadbent, D. E. (1955). Some clinical implications of recent experiments on the psychology of hearing. *Proceedings of the Royal Society of Medicine, 48*(11), 961–968.

Brown, A. L., Gjestland, T. & Dubois, D. (2016). Acoustic environments and soundscapes. In J. Kang & B. Schulte-Fortkamp (Eds.), *Soundscape and the Built Environment* (pp. 1–16). Boca Raton, FL: CRC Press.

Brumm, H., & Zollinger, S. A. (2011). The evolution of the Lombard effect: 100 years of psychoacoustic research. *Behaviour, 148*(11–13), 1173–1198.

Buchner, A., Rothermund, K., Wentura, D. & Mehl, B. (2004). Valence of distractor words increases the effects of irrelevant speech on serial recall. *Memory and Cognition, 32,* 722–731.

Cherry, E. C. (1953). Some experiments on the recognition of speech, with one and with two ears. *Journal of the Acoustical Society of America, 25,* 975–979.

Colle, H. A. & Welsh, A. (1976). Acoustic masking in primary memory. *Journal of Verbal Learning and Verbal Behavior, 15,* 17–32.

Conway, A. R. A. & Kane, M. J. (2001). Capacity, control and conflict: An individual differences perspective on attentional capture. In C. Folks & B. Gibson (Eds.), *Attraction, Distraction and Action: Multiple Perspectives on Attention Capture* (pp. 349–372). Amsterdam: Elsevier Science.

Daneman, M. & Carpenter, P. A. (1980). Individual differences in working memory and reading. *Journal of Verbal Learning and Verbal Behavior, 19,* 450–466.

Edworthy, J. (2013). Medical audible alarms: A review. *Journal of the American Medical Informatics Association, 20*(3), 584–589.

Ellermeier, W. & Zimmer, K. (1997). Individual differences in the susceptibility to the 'irrelevant speech effect'. *Journal of the Acoustical Society of America, 102*, 2191–2199.

Elliott, E. M. & Briganti, A. (2012). Investigating the role of attentional resources in the irrelevant speech effect. *Acta Psychologica, 140*(1), 64–74.

Elliott, E. M., Hughes, R. W., Briganti, A., Joseph, T. N., Marsh, J. E. & Macken, B. (2016). Distraction in verbal short-term memory: Insights from developmental differences. *Journal of Memory and Language, 88*, 39–50.

Fisher, G. G., Chaffee, D. S., Tetrick, L. E., Davalos, D. B. & Potter, G. G. (2017). Cognitive functioning, aging, and work: A review and recommendations for research and practice. *Journal of Occupational Health Psychology, 22*(3), 31.

Font-Alaminos, M. Cornella, M. Costa-Faidella, J. Hervás, A., Leung, S., Rueda, I. & Escera, C. (2020). Increased subcortical neural responses to repeating auditory stimulation in children with autism spectrum disorder. *Biological Psychology, 149*, 107807.

Gould van Praag, C. D., Garfinkel, S. N., Sparasci, O., Mees, A., Philippides, A. O., Ware, M. & Critchley, H. D. (2017). Mind-wandering and alterations to default mode network connectivity when listening to naturalistic versus artificial sounds. *Scientific Reports, 7*, 4532.

Haapakangas, A., Kankkunen, E., Hongisto, V., Virjonen, P., Oliva, D. & Keskinen, E. (2011). Effects of five speech masking sounds on performance and acoustic satisfaction: Implications for open-plan offices. *Acustica, 97*(4), 641–655.

Harvie-Clark, J. & Hinton, R. (2021, June). The value of control for acoustic satisfaction in open plan offices: A case study. In *The 13th ICBEN Congress on Noise as a Public Health Problem*, Sweden.

Hongisto, V. (2008). Effects of sound masking on workers - A case study in a landscaped office. *9th International Congress on Noise as a Public Health Problem (ICBEN)*. Foxwoods.Hongisto, V., Varjo, J., Oliva, D., Haapakangas, A. & Benway, E. (2017). Perception of water-based masking sounds-long-term experiment in an open-plan office. *Frontiers in Psychology, 8*, 1177.

Hughes, R. W. (2014). Auditory distraction: A duplex-mechanism account. *Psychology Journal, 2*, 30–41.

Hughes, R. W., Hurlstone, M., Marsh, J. E., Vachon, F. & Jones, D. M. (2013). Cognitive control of auditory distraction: Impact of task difficulty, foreknowledge, and working memory capacity supports duplex-mechanism account. *Journal of Experimental Psychology: Human, Perception, and Performance, 39*(2), 539–553.

Hughes, R. W. & Jones, D. M. (2003). Indispensable benefits and unavoidable costs of irrelevant sound for cognitive functioning. *Noise & Health, 6*, 63–76.

Hughes, R. W., Vachon, F. & Jones, D. M. (2007). Disruption of short-term memory by changing and deviant sounds: Support for a duplex-mechanism account of auditory distraction. *Journal of Experimental Psychology: Learning, Memory, and Cognition, 33*, 1050–1061.

International Organization for Standardization (2003). *ISO 354:2003* Acoustics – *Measurement* of Sound Absorption in a Reverberation Room. Geneva: International Organization for Standardization.

International Organization for Standardization (2014). ISO 12913-1:2014 *Acoustics - Soundscape - Part 1: Definition and Conceptual Framework*. Geneva: International Organization for Standardization.

International Organization for Standardization (2021). ISO 22955:2021 *Acoustics* - Acoustic Quality *of Open Office Spaces.* Geneva: International Organization for Standardization.

International WELL Building Institute (2018). *WELL* v2: *Features and Concepts.* Retrieved from: https://v2.wellcertified.com/en/wellv2/overview/

Jahncke, H., Björkeholm, P., Marsh, J. E., Odelius, J. & Sörqvist, P. (2016). Office noise: Can headphones and masking sound attenuate distraction by background speech? *Work,* 55(3), 505–513.

James, O., Delfabbro, P. & King, D. L. (2021). A comparison of psychological and work outcomes in open-plan and cellular office designs: A systematic review. *Sage Open, 11*(1), 21582440209.

Jones, D. M. (1999). The cognitive psychology of auditory distraction: The 1997 BPS Broadbent lecture. *British Journal of Psychology, 90,* 167–187.

Jones, D. M., Farrand, P. A., Stuart, G. P. & Morris, N. (1995). The functional equivalence of verbal and spatial information in short-term memory. *Journal of Experimental Psychology: Learning, Memory, and Cognition, 21,* 1008–1018.

Jones, D. M. & Macken, W. J. (1993). Irrelevant tones produce an irrelevant speech effect: Implications for phonological coding in working memory. *Journal of Experimental Psychology: Learning, Memory, and Cognition, 19,* 369–381.

Jones, D. M. & Macken, W. J. (1995). Auditory babble and cognitive efficiency: Role of number of voices and their location. *Journal of Experimental Psychology: Applied, 1,* 216–226.

Jones, D. M., Miles, C. & Page, J. (1990). Disruption of proof-reading by irrelevant speech: Effects of attention, arousal or memory? *Applied Cognitive Psychology, 4,* 89–108.

Kane, M. J. & Engle, R. W. (2003). Working-memory capacity and the control of attention: The contributions of goal neglect, response competition, and task set to Stroop interference. *Journal of Experimental Psychology: General, 132,* 47–70.

LEED (2019). *LEED v4.1 Building Design and Construction.* U.S. Green Building Council.

Leesman (2023). *The Review – Change, Issue 33.* London: Leesman. Retrieved from: https://www.leesmanindex.com/the-leesman-review/

Lenne, L. Chevret, P. & Marchand, J. (2020). *Applied Acoustics, 158,* 107049.

Leotti, L. A., Iyengar, S. S. & Ochsner, K. N. (2010). Born to choose: The origins and value of the need for control. *Trends in Cognitive Sciences, 14*(10), 457–463.

Luo, J., Wang, M. & Chen, L. (2021). The effects of using a nature-sound mobile application on psychological wellbeing and cognitive performance among university students. *Frontiers in Psychology, 12,* 699908.

Marsh, J. E., Hughes, R. W. & Jones, D. M. (2008). Auditory distraction in semantic memory: A process-based approach. *Journal of Memory and Language, 58,* 682–700.

Marsh, J. E., Hughes, R. W. & Jones, D. M. (2009). Interference by process, not content, determines semantic auditory distraction. *Cognition, 110,* 23–38.

Marsh, J. E., Yang, J., Qualter, P., Richardson, C., Perham, N., Vachon, F. & Hughes, R. W. (2018). Post-categorical auditory distraction in serial short-term memory: Insights from increased task-load and task-type. *Journal of Experimental Psychology: Learning, Memory, and Cognition, 44*(6), 882–897.

Martin, R. C., Wogalter, M. S. & Forlano, J. G. (1988). Reading comprehension in the presence of unattended speech and music. *Journal of Memory and Language, 27,* 382–398.

Miles, C., Jones, D. M. & Madden, C. A. (1991). Locus of the irrelevant speech effect in short-term memory. *Journal of Experimental Psychology: Learning, Memory, and Cognition, 17,* 578–584.

Neath, I., Farley, L. A. & Surprenant, A. M. (2003). Directly assessing the relationship between irrelevant speech and articulatory suppression. *Quarterly Journal of Experimental Psychology, 56A*, 1269–1278.

Nemecek, J. & Grandjean, E. (1973). Noise in landscaped offices. *Applied Ergonomics, 4*(1), 19–22.

Nilsson, E. (2004). Decay processes in rooms with non-diffuse sound fields Part I: Ceiling treatment with absorbing material. *Building Acoustics, 11*(1), 39–60.

Nilsson, E., Hellstrom, B., & Berthelsen, B. (2008). Room acoustical measures for open plan spaces. *Journal of the Acoustical Society of America, 123*(5), 2971.

Oseland, N. A. & Hodsman, P. (2018). A psychoacoustical approach to resolving office noise distraction. *Journal of Corporate Real Estate, 20*(4), 260–280.

Oseland, N. A. & Hodsman, P. (2020). The response to noise distraction by different personality types: An extended psychoacoustics study. *Corporate Real Estate Journal, 9*(3), 215–233.

Oseland, N. A., Tucker, M. & Wilson, H. (2022). Developing the return on workplace investment (ROWI) tool. *Corporate Real Estate Journal, 12*(2), 185–197.

Perham, N., Banbury, S. & Jones, D. M. (2007). Reduction in auditory distraction by retrieval strategy. *Memory, 15*, 465–473.

Perham, N., Begum, F. & Marsh, J. E. (2023). The categorical deviation effect may be underpinned by attentional capture: Preliminary evidenced from the incidental recognition of distracters. *Auditory Perception and Cognition, 6*(1–2), 20–51.

Perham, N. & Currie, H. (2014). Does listening to preferred music improve reading comprehension performance? *Applied Cognitive Psychology, 28*, 279–284.

Perham, N., Hodgetts, H. M. & Banbury, S. P. (2013). Mental arithmetic and non-speech office noise: An exploration of interference-by-content. *Noise and Health, 15*(62), 73–78.

Perham, N. & Macpherson, S. (2012). Mental arithmetic and irrelevant auditory number similarity disruption. *Irish Journal of Psychology, 33*(4), 181–192.

Perham, N., Marsh, J. E., Clarkson, M., Lawrence, R. & Sörqvist, P. (2016). Distraction of mental arithmetic by background speech: Further evidence for the habitual-response priming view of auditory distraction. *Experimental Psychology, 63*(3), 141–149.

Perham, N. & Sykora, M. (2012). Disliked music can be better for performance than liked music. *Applied Cognitive Psychology, 26*(4), 550–555.

Perham, N. & Vizard, J. (2010). Can preference for background music mediate the irrelevant sound effect? *Applied Cognitive Psychology, 25*(4), 625–631.

Pierrette, M., Parizet, E., Chevret, P. & Chatillon, J. (2014). Noise effect on comfort in open-space offices: development of an assessment questionnaire. *Ergonomics, 58*(1), 96–106.

Pichora-Fuller, M. K. et al. (2016). Hearing Impairment and Cognitive Energy: The Framework for Understanding Effortful Listening (FUEL). *Ear and Hearing 37*(July/August), 5S–27SRadun, J., Maula, H., Rajala, V., Scheinin, M., & Hongisto, V. (2022). Acute stress effects of impulsive noise during mental work. *Journal of Environmental Psychology, 81*, 101819.

Raghunand, S., Kameswaran, M. & Kameswaran, S. (2016). The evolution of hearing from fishes to Homo sapiens – A chronological review. *Scholars Academic Journal of Biosciences, 4*(12), 1060–1069.

Ratcliffe, E., Gatersleben, B., & Sowden, P. T. (2013). Bird sounds and their contributions to perceived attention restoration and stress recovery. *Journal of Environmental Psychology, 36*, 221–228.

Rettie, L., Potter, R. F., Brewer, G., Degno, F., Vachon, F., Hughes, R. W. & Marsh, J. E. (2023). Warning – taboo words ahead! Avoiding attentional capture by spoken taboo distractors. *Journal of Cognitive Psychology, 36*(1), 61–77.

Richardson, B., McCulloch, K. C., Ball, L. J. & Marsh, J. E. (2023). The fate of the unattended revisited: Can irrelevant speech prime the non-dominant interpretation of homophones? *Auditory Perception & Cognition, 6*(1–2), 72–96.

Röer, J. P., Bell, R. & Buchner, A. (2013). Self-relevance increases the irrelevant sound effect: Attentional disruption by one's own name. *Journal of Cognitive Psychology, 25,* 925–931.

Röer, J. P., Körner, U., Buchner, A., & Bell, R. (2017). Attentional capture by taboo words: A functional view of auditory distraction. *Emotion, 17*(4), 740–750.

Saffran, J. R., Johnson, E. K., Aslin, R. N. & Newport, E. L. (1999). Statistical learning of tone sequences by human infants and adults. *Cognition, 70*, 27–52.

Seddigh, A., Berntson, E., Jönsson, F., Danielson, C. B. & Westerlund, H. (2015). The effect of noise absorption variation in open-plan offices: A field study with a cross-over design. *Journal of Environmental Psychology, 44*, 34–44.

Smith, A. P., Jones, D. M. & Broadbent, D. E. (1981). The effects of noise on recall of categorized lists. *British Journal of Psychology 72*(3), 299–316.

Sörqvist, P. (2010). High working memory capacity attenuates the deviation effect but not the changing-state effect: Further support for the duplex-mechanism account of auditory distraction. *Memory & Cognition, 38*, 651–658.

Tremblay, S. & Jones, D. M. (1999). Change of intensity fails to produce an irrelevant sound effect: implications for the representation of unattended sound. *Journal of Experimental Psychology: Human Perception and Performance, 25*, 1005–1015.

Turner, M. L. & Engle, R. W. (1989). Is working memory capacity task dependent? *Journal of Memory and Language, 28*, 127–154.

van der Heijden, K., Rauschecker, J. P., de Gelder, B. & Formisano, E. (2019). Cortical mechanisms of spatial hearing. *Nature Reviews Neuroscience, 20*(10), 609–623.

van Meel, J. (2019). *Activity-Based Working: The Purenet Practice Guide.* Copehagen PuRE-net (The Public Real Estate Network).

Vellenga, S., Bouwhuis, T. & Höngens, T. (2017). Proposed method for measuring 'liveliness' in open plan offices. In *Proceedings of the 24th International Congress on Sound and Vibration,* London (pp. 23–27).

Waldron, S. M., Patrick, J. & Duggan, G. B. (2011). The influence of goal-state access cost on planning during problem solving. *Quarterly Journal of Experimental Psychology, 64*, 485–503.

Weber, C., Krieger, B., Häne, E., Yarker, J. & McDowall, A. (2022). Physical workplace adjustments to support neurodivergent workers: A systematic review. *Applied Psychology, 73*(3), 1–53.

7   Touch, haptics and textures

Editors' introduction

Research and lived experiences consistently indicate that textures, both those seen and those felt against our skin, have a significant effect on our thoughts and behaviours.

Even though textures can have such meaningful effects, they are regularly an afterthought, not a focus of workplace-related research or carefully considered during the design process – except to make sure that nothing will irritate the skin or be otherwise annoying.

As the material in this chapter indicates, not only do textures have powerful implications for lived experiences, but also their thoughtful selection can elevate both space user performance and wellbeing.

DOI: 10.1201/9781003390848-7

Researcher perspective

Upali Nanda

Introduction

Architecture as a profession has been defined primarily by the visual. Drawings, renderings and sketches all represent the possibility of the space, capturing the visual and often missing the other senses. The hegemony of vision makes sense given that the visual cortex is the largest part of the human brain. But this emphasis limits us in truly leveraging the power of place, especially because proximal senses like touch incite a primal reaction that can trigger emotional responses rapidly. In this chapter, we will share the research around touch and haptics, outline the importance of texture, temperature and immersion and argue for a "sensthetic" approach to designing for touch where it can enhance, or compensate, for the other senses in the modern workplace.

The untapped emotional and social potential of touch

Pallasmaa (2012) holds the ocular-centric bias and the consequent sensory imbalance responsible for the inhumanity of contemporary architecture and cities and claims that the "art of the eye" has pushed us into isolation and detachment, creating imposing and thought-provoking structures that are not rooted in humanity. In this tradition, the human body itself is abstracted to be used as a measure (proportions, units of measurement, etc.) or a metaphor (symbols, forms, etc.). As a result, modernist design has "housed the intellect and the eye but left the body and the other senses, as well as our memories and dreams, homeless" (Pallasmaa, 2012, p. 10). Merleau-Ponty and Smith (1962) argues that the manifest visibility we experience is repeated in the body by a secret visibility. Light, colour and depth, which are there before us, are there only because they awaken an echo in our body and our body welcomes them and the visible is a manifest of the invisible. In that sense, philosophically speaking, while vision is the primary sense because of all it connotes, it is also the most basic because it is necessarily connotative ... of what we touch, smell, hear, feel and so on. In fact, some theorists today place touch as the primary sense because all other senses require contact, or some form of touch, with the stimuli (food must have contact with the tongue for taste, sound waves must touch the ear, and light waves touch the eye, for hearing and sight respectively) to finally perceive.

Sitting in a workstation is a haptic and tactile experience – the feel of the chair, the typing on the keyboard, the ergonomics of posture and the comfort of the ambient temperature. As we get up and move around our kinesthetic experience changes (more information on kinesthesia follows). We feel different things, we touch different things. We craft a sensthetic of the experience beyond the visual aesthetic (Nanda, 2012).

But first, let us consider what the haptic system really is. Gibson (1966) defined the haptic system as the system responsible for perception of passive and active touch, temperature distinction, and for distinction of one's own movements – responding to skin thermoreceptors and deformation of tissues, joint configuration, stretching of muscles. It works in concert with our basic orientation system, visual system, auditory system and taste/smell system. According to Barnett (1972), touch is an essential sense because it allows "contact" with the environment through the perception of wind, humidity, temperature changes, relief, roughness, softness, etc. The comfort range of temperature for the body is between 22°C and 27°C. The internal body temperature remains at 37 degrees, while the external can range (in habitable conditions) from −1° to 46°. All climatic conditions (such as sun, wind, humidity etc.) effect the human body directly through touch. Okamoto, Nagano and Yamada (2012) identify five dimensions of touch that affect our perceptions:

1 hardness *(hard, soft)*,
2 temperature *(hot, cold)*,
3 friction properties (wet, dry, sticky, slippery),
4 fine roughness *(rough, smooth)* and
5 macro roughness (uneven, relief).

In addition to the dimensions of touch, it is important to recognise the concept of kinesthesia; our sensory awareness of the position and movement of the body which derives its meaning literally from "movement" (kinetic) and "sensitivity" (aesthesia) (Nanda, 2017). It is the sense that provides information on the whole repertory of our motor actions (Farnell, 2003). No touch, no feeling, is possible without internal and external movement – in the body and in the environment. In fact, Heller (2000) claims that touch is not a single sense. Shape perception by active touch depends crucially on combinations of inputs from the skin, movements of the finger, hand and arm in scanning and body and limb postures that afford current anchor and location cues in the absence of vision.

Recent research by Bertheaux et al. (2020) suggests that touching materials invokes an emotional response (measured by pupillary dilation) based on both value (negative/positive) and intensity. For a society, affective touch (AT) supports social connections and mitigates the effect of social conflict- a key need in today's times (Silvestri et al., 2024). Neuroimaging research shows distinct networks for AT with a key functional role for social connection (Morrison, 2016). As designers, it is also important to remember that for individuals with high sensory sensitivity, it may be perceived as less pleasant and can be subject to interindividual variability (key in discussions on neurodiversity) (Silverstri et al., 2023). That said, the importance of social touch in the context of the workplace is a key one to explore in this hybrid age.

Power of social touch and the role of in-person interactions

Social touch – the affiliative skin-to-skin contact between individuals – can rapidly evoke emotions of comfort, pleasure or calm and is essential for mental and physical wellbeing (Elias and Abdus-Saboor, 2022). As the workplace becomes an increasingly hybrid experience – there is a paucity of social touch – the hug, handshake, pat on the shoulder, etc. In fact, Suvilehto, Glerean, Dunbar, Hari and Nummenmaa (2015) argue that the degree of allowed touch can predict the closeness of a relationship – something that a remote environment does not easily allow. Thinking about touch in the workplace is not about touching or feeling inanimate objects and environments but also about touching (and feeling) human connection.

A sensthetic approach to touch, haptics and texture

Even as we isolate the sense of touch and think about perception and kinesthetics, it is important to realise that our perceptions and experiences are created by a blending of sensory perceptions- and science still cannot tell us exactly how this melding happens. In the "unity of the senses" (Marks, 2014) describes five doctrines of sensory correspondence including the first doctrine of equivalent information that postulates that different senses can inform us about the same features of the external world but in different ways. For example, movement can be perceived by sight, sound and touch, size can be perceived by sight and touch, volume can be perceived by sight, touch and sound. However, the information from different senses, about the same features, may vary perceptually. In other words, equivalence is not always proportional. For example, linear extent (or length of an object) is perceived as longer by sight than by touch. Similarly, roughness or smoothness perceived by touch is not the same as perceived by sight. In our day-to-day use of language, we often cross over sensory modalities, talking about "warm" or "cool" colours, although we sense colours visually. Our ability to pair a tactile sensation with a visual one reflects an inherent capacity to integrate or translate between modalities. In fact, metaphorically, we go back and forth between senses all the time. In a study with college students asking them about the "touch" associations of different colour students associated red, yellow and orange with "warm", blue, green and violet as "cool", and black and white as "smooth" (Nanda, 2012).

Physiologically, touch has its own "intra-modal" hierarchy; parts of the human body which are anatomically more complex have a greater area devoted to them in the somatosensory cortex, resulting in a somatosensory mapping of the external world which is out of proportion to the physical external world we "see". The tactile sensitivity of the fingers, for example, makes its role in the active exploration of environments critical (Nanda, 2012).

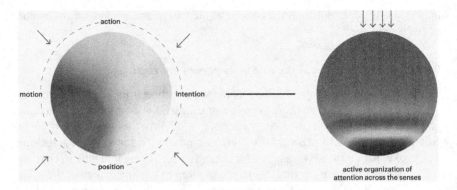

Figure 7.1 Sensthetic model of sensory and kinesthetic interaction.

To truly explore the transformative potential of touch we can lean on the "Sensthetic Model" – a conceptual tool to think "across" modalities that can help professionals and students to design, study or experience sensory environments (Nanda, 2012). The "Sensthetic Model" proposes a framework for sensory and kinesthetic interdependence for what people attend to in their environments (Figure 7.1) in terms of a dynamic and constantly re-forming organisation of senses. The underlying assumption is that the senses are plastic modalities that are fused together, with links between the different modalities. Based upon the changes in the external and internal environment there is an organisation of attention across modalities, which is a dynamic function of the intention (sum total of cognitive factors such as function, emotion, intent, and memory), motion and position (kinesthetic factors coupled to the changes in sensory modalities) and action (consequence of perception, and responsible for subsequent changes in the environment).

Note in Figure 7.1 that the colours represent the different sense modalities and arrows indicate the changes which could be in the physical or cognitive environments. Representation helps us think about touch as a sensation (and perception) closely linked to other modalities.

To truly design for tactile, haptic and kinesthetic experiences, we have to think about them in the context of all the other senses. Imagine using visuals that enhance tactile variation (through play of materials, light and shadow), or using texture to provide the precise acoustics of what we want in an environment – muted or pronounced sounds (enhancing the sense of soft or hard spaces). Our sensory mechanisms are highly interdependent, agile and plastic entities that create a beautiful opportunity for design. A few implications for practice in the workplace include the following.

1 Leaning into the array of tactile/ haptic affordances to support feelings of comfort, engagement, focus, etc.
2 Providing opportunities to physically touch each other in the workplace to develop sense of human connection (more spaces to see, hear and reach out to each other).
3 Using visual and auditory senses to enhance the "touch" of a space (soft, hard, dynamic etc.) – providing a sensorial workplace that organises attention around the primacy of touch.

Practitioner perspective

Sally Augustin

Using textures in office design

There are multiple ways that design practitioners can put textures to work, moving design-users' experiences towards desired outcomes.

- Textures that we can't touch have much less effect, so as long as the situation allows, providing opportunities for tactile contact are important. Added benefits of touch: When we touch something (and that experience is fundamentally pleasant), we're apt to come up with more creative ways for how we can use it (all of which may help up come up with a way to more effectively utilise resources provided; Kim and Krishnan, 2014).
- Soft textures, like flannel, are generally preferred to harder ones (Childers and Peck, 2010). Also popular: surfaces that are smooth, not rough (Zuo, Hope, Jones, and Castle, 2004; Wijaya, Lau, Horrocks, McGlone, Ling, and Schirmer, 2020). Designing with preferred options leads to more positive outcomes than using those are less positively evaluated.
- Having multiple pleasant tactile experiences at hand (or foot, or elbow ...) aligns with biophilic design and humans are profoundly comfortable in such spaces (Heerwagen and Gregory, 2008). Don't skimp on texture-variety. Biophilic designers also need to incorporate materials that can age gracefully over time, for example, by developing a patina, as copper and leather do (Heerwagen and Gregory). In the best case, those gracefully ageing materials are pleasant to the touch (not itchy, sticky or icky in some way, etc.).
- Smoother textures are perceived as more feminine while rougher textures are associated with masculinity (Krishna, Elder, and Caldara, 2010), which might occasionally be useful, for example, in designing a gender-specific workplace zone.
- Some materials and textures warm more rapidly from the heat of our bodies, the sun, etc., and retain that heat for longer. Using those materials and textures in social situations may be handy. A whole series of experiments have been done with people who have recently touched something warm or cool and then stopped touching that heated or cooled surface. They have found that after people have touched something warm and then stopped doing so that they find other people to be more generous and empathetic, and, in turn are more generous themselves; generally tactile warmth seems to strengthen interpersonal connections (for example, Bargh and Meinkoff, 2019; Williams and Bargh, 2008). In the same situation (i.e. having recently touched something warm) we're also

more trusting (Kang, Williams, Clark, Gray, and Bargh, 2011), which can be good or bad. Physical warmth seems to buoy up social warmth (Bargh and Shalev, 2012).

- When we sit on a cushion, even if that cushion is slim, we don't drive as hard a bargain when negotiating with others as we do when sitting on a harder surface (Ackerman, Nocera, and Bargh, 2010), we seem to interact more pleasantly. There is also some evidence that sitting on some sort of cushion encourages us to think more creatively (Xie, Lu, Wang, and Cai, 2016), making cushioned seats handy in brainstorming areas, for example. There is a place for those hard seats, however. The Xie lead team found that memory performance gets a boost when we're sitting on a harder, as opposed to a softer, surface.
- Changes of texture underfoot can be useful cues that conditions may change ahead, that there's an opening for a descending staircase, for instance. Building early warning cues into floors can save many a stumble. Similarly, differences in flooring textures, colours, or something else, can be used to speed wayfinding. People asking directions can be told to "stay on the green carpet" or to "follow the black slate path." Changes in flooring can also indicate a change in a zone – from general medicine to paediatrics, for instance. Transitions in flooring types need to be carefully managed to eliminate trip hazards.
- We tend to remain on the same sort of flooring (Bitgood, 2011). You can make it more likely that people will travel through a space in the way you would like by creating a textural path (e.g., carpeting in an otherwise uncarpeted hardwood expanse). If you want to influence the speed with which people move through a space, another of Bitgood's 2011 findings is useful: people walk more slowly on carpeted surfaces that they do on bare floors.
- Women's sense of touch is more sensitive than men's (Boles and Givens, 2011). This difference seems to be found because both men and women have about the same number of nerve endings in the ends of their fingers, etc., but since women are often shorter than men those nerve endings are closer together in women than in men. This size-based difference may, for example, explain debates among design team members and client groups about the suitability of particular textures on furniture.

In a February 2024 conversation, Meena Krenek[1] discussed using textures in interiors applications. Her experiences indicate that people have an emotion-based need to touch surfaces that they find comforting. Many textures are pleasant to touch as the body interacts with them on chair arms, underfoot, etc. Some of these positive tactile experiences are, however, also tied to prior agreeable life experiences (which may be conscious or unconsciously registered), which makes them particularly desirable project elements. Often the prior life experiences to be emulated involve natural materials, under fingertips and underfoot. Research with users identifies these upbeat associations.

---

1 Meena Krenek IIDA, ASID, LEED AP is Global Practice Director, Venues Interiors; Principal; HKS.

Krenek also reported that lighting has a significant effect on the implications of using particular textures. Light can create situations with interesting shadows and colours, for instance, while also affecting perceptions of scale and depth of field in an area.

Krenek shared that texture options can influence aesthetic balance in a space as well as apparent scale. They can generate movement and energy. Effects generated via texture can lead to effective use of data lines.

Conclusion

In summary, Krenek feels that the effects generated through texture used in a space can be similar to those created via artwork, increasing both comfort and also inspiring users both directly and via their effects on perceived depth and similar design factors, bringing users joy and comfort.

Textures felt and seen have noteworthy effects on place-user experiences. Thoughtful selections can boost space user performance and wellbeing.

References

Ackerman, J., Nocera, C. & Bargh, J. (2010). Incidental haptic sensations influence social judgments and decisions. *Science, 328*(5986), 1712–1715.

Bargh, J. & Meinkoff, D. (2019). Does physical warmth prime social warmth? Reply to Chabris et al. (2019). *Social Psychology, 50*(3), 207–210.

Bargh, J. & Shalev, I. (2012). The substitutability of physical and social warmth in daily life. *Emotion, 12*(1), 154–162.

Barnett, K. (1972). A theoretical construct of the concepts of touch as they relate to nursing. *Nursing Research, 21*, 102–110.

Bertheaux, C., Toscano, R., Fortunier, R., Roux, J. C., Charier, D., & Borg, C. (2020). Emotion measurements through the touch of materials surfaces. *Frontiers in Human Neuroscience, 13*, 455.

Bitgood, S. (2011). *Social Design in Museums: The Psychology of Visitor Studies.* Edinburgh: Museums Etc. Retrieved from: https://archive.org/details/socialdesigninmu0000bitg.

Boles, D. & Givens, S. (2011). Laterality and sex differences in tactile detection and two-point thresholds modified by body surface area and body fat ratio. *Somatosensory and Motor Research, 28*(3/4), 102–109.

Childers, T. & Peck, J. (2010). Informational and affective influences of haptics on product evaluation: Is what I say how I feel?" In A. Krishna (ed.), *Sensory Marketing: Research on the Sensuality of Products.* New York: Routledge, pp. 63–72.

Elias, L. J. & Abdus-Saboor, I. (2022). Bridging skin, brain, and behaviour to understand pleasurable social touch. *Current Opinion in Neurobiology, 73*, 102527.

Farnell, B. (2003). Kinesthetic sense and dynamically embodied action. *Journal for the Anthropological Study of Human Movement, 12*(4), 132.

Gibson, J. J. (1966). *The Senses Considered as Perceptual Systems.* London: Bloomsbury Academic.

Heerwagen, J. & Gregory, B. (2008). Biophilia and sensory aesthetics. In S. Kellert, J. Heerwagen & M. Mador (eds.), *Biophilic Design.* Hoboken, NJ: John Wiley and Sons, pp. 227–241.

Heller, M. A. (2000). *Touch, Representation and Blindness*. Oxford, UK: Oxford University Press.

Kang, Y., Williams, L., Clark, M., Gray, J. & Bargh, J. (2011). Physical temperature effects on trust behavior: The role of insula. *Social Cognitive and Affective Neuroscience, 6*(4), 507–515.

Kim, H. & Krishnan, S. (2014). Where is the fun in creativity? The influence of product touch on consumer creativity. *Abstracts, Conference of the Society for Consumer Psychology,* March 6–8, Miami.

Krishna, A., Elder, R. & Caldara, C. (2010). Feminine to smell but masculine to touch? Multisensory congruence and its effect on the aesthetic experience. *Journal of Consumer Psychology, 20*(4), 410–418.

Marks, L. E. (2014). *The Unity of The Senses: Interrelations among the Modalities*. New York: Academic Press.

Merleau-Ponty, M. & Smith, C. (1962). *Phenomenology of Perception* (Vol. 26). London: Routledge.

Morrison, I. (2016). Affective and social touch. In J. D. Greene, I. Morrison & M. E. P. Seligman (eds.), *Positive Neuroscience*. New York: Oxford Academic.

Nanda, U. (2012). *Sensthetics: A Crossmodal Approach to Sensory Design*. Saarbrücken, Saarland: AV Akademikerverlag.

Nanda, U. (2017). A sensthetic approach to designing for health. *Journal of Interior Design, 42*(2), 7–12.

Okamoto, S., Nagano, H. & Yamada, Y. (2012). Psychophysical dimensions of tactile perception of textures. *IEEE Transactions on Haptics, 6*, 81–93.

Pallasmaa, J. (2012). *The Eyes of the Skin: Architecture and the Senses*. Chichester: John Wiley & Sons.

Silvestri, V., Giraud, M., Macchi Cassia, V., & Nava, E. (2024). Touch me or touch me not: Emotion regulation by affective touch in human adults. *Emotion, 24*(4), 913–922.

Suvilehto, J. T., Glerean, E., Dunbar, R. I. M., Hari, R. & Nummenmaa, L. (2015). Topography of social touching depends on emotional bonds between humans. *Proceedings of the National Academy of Sciences of the United States of America, 112*, 13811–13816.

Wijaya, M., Lau, D., Horrocks, S., McGlone, F., Ling, H. & Schirmer, A. (2020). The human 'feel' of touch contributes to its perceived pleasantness. *Journal of Experimental Psychology: Human Perception and Performance, 46*(2), 155–171.

Williams, L. & Bargh, J. (2008). Experiencing physical warmth promotes interpersonal warmth. *Science, 322*(5901), 606–607.

Xie, J., Lu, Z., Wang, E. & Cai, Z. (2016). Remember hard but think softly: Metaphorical effects of hardness/softness on cognitive functions. *Frontiers in Psychology, 7*, article 1343.

Zuo, H., Hope, T., Jones, M. & Castle, P. (2004). Sensory interaction with materials. In D. McDonagh, P. Hekkert, J. van Erp & D. Gyi (eds.), *Design and Emotion: The Experience of Everyday Things*. New York: Taylor & Francis, pp. 223–227.

8   Promoting healthy behaviours

Editors' introduction

Creating and maintaining mentally and physically healthy workplaces is the loftiest mission of workplace researchers and designers, one that they often feel is thwarted by businesses' focus on profit. However, worker health and wellbeing and organisational productivity and success are not mutually exclusive endeavours, as this chapter in this book makes clear. Workplace design that optimises user performance is a design that elevates their wellbeing as well.

When our subjective cognitive wellbeing goes up, Armenta, Ruberton, and Lyubomirsky (2015) report that that 'leads to a variety of beneficial outcomes via an increase in behaviours that offer individuals the opportunity to achieve success in multiple domains'. The Armenta team shares that when people have greater subjective wellbeing they thrive socially, establishing successful relationship with others, for example, they're also able to successfully cope with life changes, more likely to be satisfied with their jobs (and their supervisors report that their professional performance is better), as well as being more creative and dependable. These individuals are also likely to be in better health than individuals with lower wellbeing levels.

Research also consistently shows a link between mood, wellbeing, and physical health; as the mental state is elevated, physical processes follow (Segerstron and Sephton, 2010; Sternberg, 2009). Physical stressors, such as temperatures that are too high or too low, or distracting noise, etc., can have particularly deleterious effects on mood and wellbeing. Clearly, there's more to creating a workplace that promotes healthy behaviours, where people are healthy, than fiddling with the ventilation.

DOI: 10.1201/9781003390848-8

Researcher perspective

Susanne Colenberg and Jos Kraal

Introduction

The first part of this research section defines healthy behaviour in the workplace and explains the mechanisms of behavioural change through interior design. The second part discusses the available evidence on the health impact of workplace design elements, such as furniture, layout, and visual prompts. It concludes with an overview of current knowledge on promoting healthy behaviours at the office.

Workplace design approaches to impact health behaviour

How to design an office that supports a healthy lifestyle? According to the World Health Organisation, health equals wellbeing and has a physical, psychological, and social dimension (WHO, 2006). This implies that a healthy office should encourage behaviours that support, stimulate, and maintain all three wellbeing dimensions and discourage behaviours that undermine one or more of these dimensions. When determining which behaviours the workplace design should encourage or discourage, it is important to note that wellbeing has two components: an objective, observable, or diagnosable component, such as a person's medical state, and a subjective, perceived component, such as mood or satisfaction. Both objective and subjective wellbeing include short-term experiences or snapshots that fluctuate over time and more stable long-term states. When designing and evaluating workplaces that intend to promote employee wellbeing, these components and states need to be taken into account.

After defining the desired behaviours and analysing current obstacles, a design strategy can be developed. For starters, it has to be possible for the users to perform the desired behaviour in the new environment. In other words, the design should *afford* the target behaviour. According to the ecological psychologist Gibson (1977), the physical environment is composed of surfaces and objects that jointly enable or disable user activities. He called these action possibilities 'affordances'. For example, if the employees should be encouraged to take the stairs instead of the elevator, at least the office should afford the action of stair walking by featuring accessible and walkable stairs.

Although Gibson was convinced that affordances are intuitively understood by users, in practice not all affordances may be perceived as such due to limitations in physical or mental capacity, social conventions, or specific circumstances. For example, employees with walking difficulties, visual impairments, or vertigo may perceive barriers to using stairs. Elevators are perceived as quick and effortless, and if there is a hurry to bring something from the 3rd to the 10th floor, only trained athletes will consider using the stairs instead of the elevator. Furthermore, the context may signal the inappropriateness of the use of spaces or furniture. If staircases look like they should only be used in case of an emergency, employees will be hesitant to use them daily. Similarly, organisational culture may influence the perception of affordances and make behaviours that are technically possible a less desired option.

The characteristics of a space that communicates behavioural options may be considered psychological affordances and are at a higher abstraction level than functional affordances that technically enable action (Colenberg et al., 2023). To promote healthy behaviour, both functional and psychological affordances are required and should be aligned. The users have to understand the behavioural setting and feel free, capable, and invited to perform the behaviour. Therefore, it is of imminent importance to identify the different user groups of the workplace and learn about their capabilities, backgrounds, preferences, and expectations and use this knowledge as input for the design. The COM-B model (Michie et al., 2011), a summary of several behavioural theories, may be a useful framework to analyse the capacity, opportunity, and motivation of the users to perform the target behaviour.

In addition to making it physically and psychologically possible to perform healthy behaviour, the workplace design could actively stimulate the behaviour through nudging. The concept of nudging refers to behavioural techniques to guide people in the desired direction by interfering in their unconscious decision processes and gently suggesting a specific choice (Thaler and Sunstein, 2008). The British Behavioural Insights Team has developed a pragmatic framework, EAST, which points out that nudges are most effective when they are Easy, Attractive, Social, and Timely (Service et al., 2015). This framework can be a useful starting point for designers thinking about applying nudges to promote healthy behaviour in the workplace.

Structural nudges that are integrated into the physical environment are often more effective than one-off nudges (Van Woerkom, 2021). This may be because long-term exposure to these nudges can create new habits (Aarts and Dijksterhuis, 2000). Indeed, interventions at the workstation have a more sustainable effect than educational interventions (Zhu et al., 2020). Additionally, nudges work best when they align with the users' intentions, such as adhering to a healthy lifestyle, or in situations where they experience conflicting preferences (Venema and van Gestel, 2021), which may well apply to healthy behaviours at the office. Regarding workplace design, nudges may include both spatial and decorative elements and persuasive technology that is incorporated into the architectural design, such as dynamic decoration or sensors that provide a reward (e.g. music or applause) when the desired behaviour is detected. However, the long-term effectiveness of nudges often is not clear and there are ethical concerns in applying non-transparent nudges.

Promoting physical activity at the office

The peer-reviewed research on promoting healthy behaviour at the office is limited; it largely focuses on improving physical wellbeing by reducing sedentary behaviour and increasing physical activity (Colenberg and Jylhä, 2022). Incorporating physical activity into daily life through thoughtful building design is referred to as active design (Engelen, 2020). Because the office workplace is recognised as an environment where people spend extensive periods of time sitting, applying active design in this context may significantly impact employee health. After all, prolonged sitting is associated with obesity, diabetes, cardiovascular disease, and premature mortality, and breaking up sedentary time reduces these risks (Dunstan et al., 2011; Neuhaus et al., 2014).

The most studied intervention to decrease sitting time is the implementation of sit-stand desks whose height can be adjusted to work in both sitting and standing positions. In general, sit-stand desks reduce sitting and increase standing time (Neuhaus et al., 2014; Zhu et al., 2020) but there may be comfort issues (Karakolis and Callaghan, 2014) and the effect may diminish over time. Venema et al. (2018) showed that setting a sit-stand desk by default at standing height substantially increases stand-up working. The use of treadmill desks with bicycle pedals underneath can increase physical activity (Zhu et al., 2020) and decrease body fat (Torbeyns et al., 2016).

Apart from providing activating furniture or activity-permissive workstations, the physical activity of employees can be influenced by the office's layout. To increase the frequency of non-sedentary breaks, a designer can opt for a larger variety of walking routes, referred to as local connectivity of the workspace, and greater proximity and visibility of co-workers (Duncan et al., 2015; Wilkerson et al., 2018). Furthermore, office workers spend less time seated in an activity-based working environment, which offers a variety of workspaces designed to support specific work activities, than in a traditional office environment (Foley et al., 2016). However, offering an appealing and easily accessible staircase, breakout spaces, and centralised facilities can reduce sitting time but may not increase moderate or vigorous physical activity (Jancey et al., 2016). Increased distances between the workspaces and communal facilities, such as the bathroom and kitchen, do not seem to lead to more walking (Engelen et al., 2016, 2017; Sawyer et al., 2017). Therefore, it may be best to combine these affordances with other strategies to promote physical activity at the office.

Additional strategies could include nudging stair use, an activity that can reduce cardiovascular disease risk (Meyer et al., 2010). Different nudging strategies have been tested with mixed results. In a field experiment by Swenson and Siegel (2013), an interactive artwork located within the staircase of a three-storey office and additional signs near the staircase's entrance doubled the stair usage. This effect lasted for at least six weeks. In a study by Ferrara and Murphy (2013), motivational signs were more effective in promoting stair use than art murals on the staircase. Whereas Moloughney et al. (2019) showed that enhancing the stairwell with wall paint, upgraded stair treads and handrails, artwork, and glass doors increased stair usage in the long term without additional prompts.

A review by Nocon et al. (2010) indicates that the effects of prompts such as posters, floor stickers, and stair banners often are inconsistent or non-significant. In a study by Lewis and Eves (2012), point-of-choice prompts were effective but motivational posters within elevators showed no positive results. In another study, posters initially boosted stair usage, but this dropped back to the baseline after their removal (Kwak et al., 2007). And, in one instance, the applied nudges resulted in reduced stair usage because they annoyed the occupants of the office (Åvitsland et al., 2017). This underlines that prompts can be effective nudges when applied in the right format, time, and location, and when they are embraced by the target audience. This aligns with the aspects of the previously mentioned EAST framework (Service et al., 2015).

At the office, the stairs usually have to compete with the elevators, and therefore both have to be considered in office design. In a natural experiment, Nicoll and Zimring (2009) studied stair use in an office building that featured two types of elevators: a skip-stop elevator that only stopped at every third floor and a traditional elevator that stopped at every floor. The users of the skip-stop elevator were expected to walk up or down on nearby, visible, and attractive stairs. The employees located near the skip-stop elevators used the stairs 33 times more than the employees near the traditional elevator. However, the presence of open and central staircases in other office buildings did not result in increased walking (Engelen et al., 2016, 2017).

Additionally to promoting physical activity within the office building, active design could involve promoting active commuting (walking, biking) by offering facilities at the office for safe and easy bicycle parking (Zhu et al., 2020), charging e-bikes, bicycle maintenance, showering, changing clothes, and drying gear. Commuting by car may be discouraged by reducing the number of car parking spots. Note that in workplaces other than offices, it may be beneficial to employees' health to reduce rather than increase physical activity at work.

Stimulating healthy food choices

Strategies to impact healthy food choices have been widely studied (Arno and Thomas, 2016). However, these studies nearly all focus on the adjustment of portion sizes, packaging, ordering processes, or informing users about nutrients and seldomly address the physical work environment. Nevertheless, studies in retail environments indicate that aspects like visibility, positioning, and accessibility can increase healthy food choices, such as the placement of healthy items near the checkout (Cheung et al., 2019; Kroese et al., 2016). These insights may be relevant to the layout of the office cafeteria and the design of the food counters.

Research shows that convenience is an important driver for healthy food choices. German scholars found that a malfunctioning all-inclusive buffet led to long waiting lines, which made employees switch to the healthy food counter (Bauer et al., 2021). However, dietary behaviour quickly returned to baseline levels after the all-inclusive terminal was fixed. Adding green footsteps towards, the healthy food counter had virtually no effect. Interestingly, the researchers had abandoned their intention to pair the footsteps with health- and dieting-related priming words because they received negative feedback on their healthy food campaign. Again, this indicates that nudging needs careful fine-tuning to the attitude of the target group to be effective. Apparently, for this group, avoiding hassle was a stronger motivation for food choice than increasing health and paternalistic reminders caused resistance. Furthermore, nudges may be less effective when people have strong routines. In a Danish hospital, a chef's recommendation sticker and prominent positioning of vegetarian sandwiches increased their purchase by visitors but not by staff (Venema and Jensen, 2023).

Improving mental and social wellbeing

In contrast to the research on increasing physical wellbeing through workplace design, research on interior design strategies to promote behaviours that increase mental or social wellbeing is scarce. Usually, mental wellbeing in the workplace is promoted through training and education, therapy techniques, e-health, and wellness programmes rather than interventions in the physical environment. Mental health-supporting behaviours like taking breaks, and immersing yourself in focused work, may be promoted by the presence of restorative spaces, relaxing chairs, private workspaces, quiet-working zones, visibility of breakout spaces, and prompts that remind employees of taking breaks, caring for plants, or engaging in explorative activities and learning new things. In healthcare environments, positive wellbeing effects have been found of 'energy pods', cabins or chairs for short restorative naps (Dore et al., 2021).

To promote employees' social wellbeing, the workplace design should afford identity expression, enable social interaction, and provide privacy (Colenberg, 2023; Spreitzer et al., 2020). This may include nudging employees to, for example, customise their environment and invite them to have a chat or remind them to be quiet in workspaces for focused work. Centralised and well-connected spaces are used more intensively and attract more visitors, which may increase spontaneous encounters and afford human connections (Sailer and Koutsolampros, 2021).

Olsson et al. (2020) present several design solutions for actively facilitating, inviting, and encouraging social interactions between collocated people using technology, for example, ice-breaking games and interactive installations. The design solutions aim to improve the quality, value, or extent of social interaction by, for example, increasing awareness of other people in one's surroundings, nurturing ongoing interactions, supporting a sense of community by revealing common ground, or engaging people in collective activity. Their examples of designs related to interior space include an interactive floor, an interactive installation that displays the overall moods of the participating employees in a light pattern projected in a hallway, a tabletop videogame that starts when a coffee mug touches the table, and a display that presents photos in the online gallery of users who are close by. Unfortunately, research on their effect on social interactions was limited.

Conclusion

Workplace design may have a significant and enduring impact on employees' behaviour that includes activities that support and maintain their physical, mental, and social wellbeing. The current research includes several examples of affordances and nudges that have been found to promote physical activity and healthy food choices. However, evidence-based design solutions for promoting mental wellbeing and enhancing social relationships are lacking. Expertise in behaviour change, decision-making and a human-centred design approach are required to make a structural impact on the employees' health behaviour.

Practitioner perspective

Deborah Bucci

A practitioner's dilemma

It is very, very, very, very hard to change human behaviour.
Ron Goetzel, Johns Hopkins Bloomberg
School of Public Health (2018)

The realities of the workplace in the 21st century present a shift from the industrial age of the 18th to the early 20th centuries, where the focus on workplace design was solely on productivity and efficiency, often at the cost of human wellbeing, into the human era when workplace design is supposed to prioritise employee wellbeing, health, and overall satisfaction in hopes of improving productivity. Safety is no longer the only focus in the workplace; there is a growing focus on health promotion and disease prevention. Wellness programmes and WELL Buildings are the approaches used most often to develop physical and virtual workplaces that boost user mental and physical health.

Even considering all that is known about behaviour change, organisations are still trying to 'practice wellness' by putting the onus on the individual to pivot away from unhealthy behaviours through programmes and incentives without taking any responsibility for unhealthy workplace environments. This approach is ineffective and costly (Miller et al., 2018). Workplace wellness programmes to promote healthy behaviours have been around since the 1970s (Lewis, 2012). These organisation-sponsored programmes have offered a plethora of goods and services to encourage employees to move more, eat healthier, and reduce stress, all to keep employees engaged and manage healthcare costs for those without universal coverage. The wellness industry is a booming business, with 1.8 trillion dollars in sales annually, yet the evidence for the effectiveness of these programmes is sparse (Global Wellness Institute, 2024). Today's workers are experiencing high burnout and mental distress and continue to suffer from chronic disease at an alarming rate (Gallup, 2024). Engagement and productivity of workers continue to decline, perplexing organisational leadership while consultants and Human Resource departments continue to search for the miracle cure.

Colenberg and Kraal referenced the World Health Organisation's definition of health, which comprises three key components: physical, psychological, and social dimensions that contribute to the wellbeing of an individual. Further, wellbeing can be viewed from an objective and subjective perspective, which can be multifactorial, elusive, and open to many interpretations.

Defining wellbeing is an elusive endeavour. The numerous vantage points to view the concept add to the complexity. The words wellness and wellbeing are so often interchanged that one can wonder which is being used in the literature. Individuals, organisations, healthcare, insurance, academia, industry, retail, and government, to name a few, all weigh in with interpretations, metrics, and standards. Who gets to decide the working definition?

What do workplace wellness programmes do? Evidence from the Illinois workplace wellness study. A study supported by the National Institutes of Health (NIH) and National Science Foundation (NSF) concluded that there were no significant effects of wellness programmes on measured outcomes of healthcare spending, employee productivity, and health behaviours. A few specifics were that higher incentives got people started but people did not remain in programmes, healthier people self-selected into wellness programmes, and unhealthy people were likely not to participate.

> After one year, we find no significant effects of our wellness program on the many outcomes we examine, with two exceptions: employees are more likely to have received a health screening and to believe that the employer places a priority on worker health and safety
>
> (Jones et al., 2019)

When the programmes are not working and the individual workers are not changing their behaviours, the next place to focus is on the physical workplace environment.

> If I led a company with a lot of employees, I would spend money on environment rather than spend money telling them to change their bad behaviours. If someone's environment is going to dramatically impact their health and productivity – that is where I would focus.
>
> Al Lewis (2012)

Why nobody believes the numbers: distinguishing fact from fiction in population health management

As Susanne Colenberg Jos Kraal mentioned, before recommendations can be levied to support promoting healthy behaviours in the workplace, it *could* be essential to identify, define, and determine the gap between the current state and desired outcomes. Additionally, it might be necessary to discuss who the workers are, the industries/occupations in which they are employed, and how their work impacts wellbeing. Further, what do employees need to do their best work? Within each context, workers experience different stressors; some have to do with how they work, some have to do with what type of work they are performing, or a combination thereof, hence requiring a multitude of different interventions to obviate the effects on their mind, body, or spirit. There is no a one-size-fits-all solution.

Does design need to *promote* healthy behaviours, facilitate behaviour change, or design a healthy workplace where any space inhabitants will benefit by just being there? Kate Lister and Tom Harnish wrote, 'The whole person, not just the "employee," comes to work each day and goes home each night' (Miller et al., 2018).

They carry into and out of work all the complexities of their personal life circumstance along with other worries, fears, challenges, frustrations, hopes, and dreams. Wouldn't it make sense to design a building to support the wellbeing of everyone, no matter their circumstance or reason for being there?

The onus on the building industry

Everyone benefits if buildings are designed to be healthy and advocate for the wellbeing of all. The WHO laid out some ground rules for the minimum standards in the physical work environment. The most comprehensive endeavour towards prioritising health and wellness in the built environment is the WELL Building Standard, launched in 2014 by Delos[1] and administered by the International WELL Building Institute, a subsidiary of Delos. This evidence-based programme includes ten critical areas focused on every aspect of environmental design, and recommendations extend beyond the built environment. The ten key areas are air, water, nourishment, light, movement, thermal comfort, sound, materials, mind, and community. Paul Scialla, founder of Delos, posits that humans spend 90% of their time indoors, between four walls and a roof. 'What if we could activate that space to provide a passive and constant delivery of preventative medical benefits that would not require the occupant to do anything…' (Miller et al., 2018).

Colenberg and Kraal highlight recommendations for specific design elements that fall under some of the critical areas noted by Delos. It is important to reiterate that context matters. Nudges were also introduced as a method of encouraging healthy behaviours. The effectiveness of nudges in the workplace depends on a combination of factors, especially context. When carefully designed, ethically implemented, and aligned with organisational goals, nudges can be valuable for influencing positive behaviours and promoting a conducive work environment.

Case study

A global media company had just completed a massive upgrade to one of its campuses in anticipation of a future relocation of employees from other sites in the area to this one central campus. An urban, edgy design was carried through all the buildings on campus. The colour scheme used for decor and furniture was on trend with design elements to enhance creativity, collaboration, calmness, and focus. Exciting art and sculpture were scattered between the corporation's long-history relics. Comfy, inviting furniture was placed strategically throughout the buildings, offering spaces for solo work and collaborative meetups. Quiet, meeting, and larger conference rooms dotted the campus, as did a few phone booths. Each division had a hydration station, eating space, food storage, and warming equipment. A state-of-the-art dining area with a marketplace design that offered every type of cuisine imaginable was centrally located and available 24/7. The main entrance to the campus housed a Starbucks with plenty of seating. The overall interior design

1   https://delos.com/

capitalised on capturing natural light, offering the opportunity for numerous plants throughout the campus. An onsite gym with fitness classes, locker rooms, and table tennis areas was available to employees for a nominal fee.

Numerous outdoor spaces with tables and chairs, high-speed internet, fire pits, and hammocks offered places to gather and work. Walking trails through gardens with water features connected the parking garage to the buildings, with plenty of bicycle parking in a protected shelter. Employees would pass through security and be lured into the inviting space.

The global media company was dedicated to supporting employee wellbeing. Unsurprisingly, when there was an increase in stress-related health claims and consistent feedback from employees reporting stress-related health issues (insomnia, headaches, IBS etc.) from their onsite healthcare centre, it peaked concern, and the organisation decided to look at what they could do to improve wellbeing in the workforce.

The leadership brainstormed some strategies based on best practices in other organisations, current literature, and experts in the field. It was decided to run a pilot study utilising their top three choices; however, before implementing any initiatives, a brief two-question pre-pilot survey was sent to get employee feedback. One hundred and five employees participated in a pre-pilot survey that asked two questions; those questions and their most frequent responses follow:

- What do you want help with most?
  - Manage stress and reduction.
  - Improve cognitive skills.
  - Increase creativity.
- What are your most significant barriers?
  - Finding the time.
  - Not knowing where to start.
  - My work environment is not conducive.

The overwhelming responses to both questions were quite telling. The workplace was not conducive to cognition or creativity, which caused employees to experience various types of stress. Combined with employees' time crunch and lack of knowledge of a clear pathway forward, it became clear why employees were suffering.

With this information at hand, a pilot was created, which offered an employee the opportunity to attend mindful meditation classes and yoga, a free subscription to a meditation app, and the opportunity to attend a Corporate Athlete Course that was offered on campus. What about changes related to workplace design?

Proceedings from a corporate athlete course

The day begins with introductions, roles at the organisation, and expectations. Below are a few responses generated by this group of attendees:

1 Learn how to have more energy at the end of the day – I am so beaten down when I leave.

2 Learn how to recover from stress in the moment and throughout the day and how to do that in my cube.
3 Wellness refresh. I do not know how to take care of myself at work anymore, and I do not know if that is valued.
4 I work all the time because I cannot concentrate in the open workspace. I need to learn tips to focus better.
5 Energy. I need more energy. I am taking care of my terminally ill mom while trying to work and care for my family.
6 I have been moved into an open workspace. I have no privacy, and the distraction is incredible. I cannot concentrate at work – so I must take my work home. So, I work all day. How can I get better focused during the day?
7 We have this great gym here, and I want to work out daily; however, I get much flak about going. When I do, I come back so focused. I want to be able to do that freely without being judged. How can I help change the mindset and culture?

After so much effort in designing the campus, the comments relating to the impact of the workspace design on employee wellbeing came as a surprise. Leadership took the following actions:

1 Communicated the commitment to employee wellbeing and added an employee representative to their design team.
2 All spaces on campus were available for employee use in whatever way it served them.
3 Several conference rooms were converted into quiet rooms and furnished with comfortable furniture for rest breaks and focused work on a drop-in basis.
4 A wellness break room for naps (by reservation) was created.
5 Two meditation booths were placed in quiet and accessible spaces on campus.
6 While converting the open desk spaces entirely was impossible, smaller work pods with less density were created.
7 Healthy snacks and fresh fruit were available in all break rooms.

The pilot became an enterprise-wide initiative offering these programmes domestically and in several international locations. Adding an employee to the design team provided continuous feedback, which was valuable as more teams relocated to the central campus.

Discussion

How can I help change the mindset and culture?

This comment sums up what employees were experiencing: A disconnect between what they needed to do their best work and the environmental support offered through building design. Despite the incredible architectural design, employees still experienced a loss of wellbeing by being in the space. There was a stigma around using the environment and its amenities. It is unclear how much input employees had in the design decisions. What was clear is that the workspaces offered to these employees did not support their needs for focus, privacy, and autonomy during their workday. What could help?

'Holding space' is often used in therapeutic and personal development settings. The meaning generally refers to offering support, empathy, and a non-judgemental presence to someone going through a challenging time. What if the workplace design could create physical and environmental conditions that support employees' wellbeing, collaboration, and productivity by creating a physical environment that promotes safety, comfort, and inclusivity? In essence, holding space for whatever might happen within the four walls on any given day.

Here's how the concept may apply to workplace design:

- *Physical Comfort* – Providing ergonomic furniture, comfortable seating, and adjustable workstations can contribute to physical wellbeing. Ensuring proper lighting, ventilation, and acoustics also plays a role in creating a comfortable workspace.
- *Flexibility and Adaptability* – A well-designed workplace allows for flexibility and adaptability. This includes versatile workspaces that can be easily reconfigured to accommodate different tasks, collaboration, and individual work.
- *Inclusive Design* – Designing spaces that are inclusive and considerate of diverse needs fosters a sense of belonging. This can involve providing a variety of spaces for different work styles, preferences, and accessibility requirements.
- *Biophilic Design* – Incorporating elements of nature, such as plants and natural light, can positively impact wellbeing. Biophilic design principles aim to connect people with nature within the built environment, promoting a healthier and more pleasant atmosphere.
- *Noise Management* – Addressing noise concerns through the strategic placement of quiet zones, sound-absorbing materials, or dedicated collaborative spaces helps create an environment conducive to concentration and collaboration.
- *Technology Integration* – Implementing technology solutions that enhance productivity and reduce stress can contribute to holding space in the workplace. This includes tools for efficient communication, task management, and a seamless work experience.
- *Wellness Rooms* – Providing designated wellness or relaxation spaces allows employees to take breaks, practice mindfulness, or engage in activities that support mental and emotional wellbeing.
- *Social Spaces* – Designing communal areas where employees can connect, collaborate, and build relationships helps foster a positive workplace culture. This can include breakout areas, cafeterias, or collaborative workspaces.
- *Personalisation and Control* – Allowing employees to personalise their workspaces and providing control over environmental factors (such as temperature and lighting) can contribute to a sense of ownership and wellbeing.
- *Supportive Leadership and Policies* –'Holding space' in the workplace extends beyond the physical environment. It also involves leadership practices and policies that support employees' wellbeing, work-life balance, and mental health.

Conclusion

Promoting healthy behaviour through workplace design is best accomplished by two key elements. First and foremost, workplace design should benefit all, encompassing the elements listed by the WELL Building Standard that support physical, mental, and social wellbeing.

The second crucial factor is creating a culture of care in the organisation. While designers cannot change or fix a toxic work culture, they can ask deep questions about the intended use of the physical space and, in a sense, nudge a focus on worker wellbeing. Leadership and stakeholders must be committed and consider worker wellbeing in all strategic decisions, which builds a health-focused culture. Workers should be involved in every step of the design process, from initial inquiry to follow-up evaluations. An analysis of the gap from the current state of wellbeing into an ideal, and considering if it is attainable, is crucial. Organisations must look at best practices and talk with subject matter experts, academics, and practitioners to assemble a complete picture of what could increase mental and physical health in their offices. Design solutions need to be sustainable through enmeshment and integration into workplace culture. These solutions and strategies cannot exist in a silo. There needs to be evaluation and continuous improvement to keep up with the dynamic changes of the world.

References

Aarts, H. & Dijksterhuis, A. (2000). Habits as knowledge structures: Automaticity in goal-directed behavior. *Journal of Personality and Social Psychology*, *78*(1), 53–63.

Arno, A. & Thomas, S. (2016). The efficacy of nudge theory strategies in influencing adult dietary behaviour: A systematic review and meta-analysis. *BMC Public Health*, *16*(1), 676.

Åvitsland, A., Solbraa, A. K. & Riiser, A. (2017). Promoting workplace stair climbing: Sometimes, not interfering is the best. *Archives of Public Health*, *75*(2).

Bauer, J. M., Bietz, S., Rauber, J. & Reisch, L. A. (2021). Nudging healthier food choices in a cafeteria setting: A sequential multi-intervention field study. *Appetite*, *160*, 105106.

Cheung, T. T. L., Gillebaart, M., Kroese, F. M., Marchiori, D., Fennis, B. M. & De Ridder, D. T. D. (2019). Cueing healthier alternatives for take-away: A field experiment on the effects of (disclosing) three nudges on food choices. *BMC Public Health*, *19*(1), 1–10.

Colenberg, S. (2023). *Beyond the Coffee Corner: Workplace Design and Social Wellbeing.* Delft University of Technology.

Colenberg, S., Appel-Meulenbroek, R., Romero Herrera, N. & Keyson, D. (2023). Interior designers' strategies for creating social office space. *Ergonomics, 67*(7), 886–1007.

Colenberg, S. & Jylhä, T. (2022). Identifying interior design strategies for healthy workplaces: A literature review. *Journal of Corporate Real Estate*, *24*(3), 1463–1464.

Dore, E., Guerero, D., Wallbridge, T., Holden, A., Anwar, M., Eastaugh, A., Desai, D. & Clare, S. (2021). Sleep is the best medicine: How rest facilities and EnergyPods can improve staff wellbeing. *Future Healthcare Journal*, *8*(3), e625.

Duncan, M. J., Short, C., Rashid, M., Cutumisu, N., Vandelanotte, C. & Plotnikoff, R. C. (2015). Identifying correlates of breaks in occupational sitting: A cross-sectional study. *Building Research and Information*, *43*(5), 646–658.

Dunstan, D. W., Thorp, A. A. & Healy, G. N. (2011). Prolonged sitting: Is it a distinct coronary heart disease risk factor? *Current Opinion in Cardiology*, *26*(5), 412–419.

Engelen, L. (2020). Does active design influence activity, sitting, wellbeing and productivity in the workplace? A systematic review. *International Journal of Environmental Research and Public Health*, *17*(24).

Engelen, L., Chau, J., Bohn-Goldbaum, E., Young, S., Hespe, D. & Bauman, A. (2017). Is active design changing the workplace? A natural pre-post experiment looking at health behaviour and workplace perceptions. *Work*, *56*, 229–237.

Engelen, L., Dhillon, H. M., Chau, J. Y., Hespe, D. & Bauman, A. E. (2016). Do active design buildings change health behaviour and workplace perceptions? *Occupational Medicine*, *66*(5), 408–411.

Ferrara, C. M. & Murphy, D. (2013). Motivational signs, artwork, and stair use in a university building. *Californian Journal of Health Promotion*, *11*(1), 76–83.

Foley, B., Engelen, L., Gale, J., Bauman, A. & Mackey, M. (2016). Sedentary behavior and musculoskeletal discomfort are reduced when office workers trial an activity-based work environment. *Journal of Occupational and Environmental Medicine*, *58*(9), 924–931.

Gallup, Inc (2024). *The State of the Global Workplace*. Retrieved from: https://www.gallup.com/workplace/349484/state-of-the-global-workplace.aspx

Goetzel, R., (2018). Johns Hopkins Bloomberg School of Public Health. https://coeindy.com/the-invisible-power-of-the-workplace/

Gibson, J. J. (1977). The theory of affordances. In R. E. Shaw and J. Bransford (Eds.), *The Ecological Approach to Visual Perception* (pp. 67–82). Lawrence Erlbaum.

Global Wellness Institute (2024). *The Global Wellness Economy Country Rankings*. Retrieved from: https://globalwellnessinstitute.org/industry-research/2024-the-global-wellness-economy-country-rankings/

International WELL Building Institute (2020). *WELL v2*. Retrieved from: https://v2.wellcertified.com/en

Jancey, J., Sarah McGann, Robyn Creagh, Krysten Blackford, Peter Howat, and Marian Tye. 2016. "Workplace building design and office-based workers' activity: A study of a natural experiment." *Australian and New Zealand Journal of Public Health*, *40*(1),78–82.

Jones, D., Molitor, D. & Reif, J. (2019). What do workplace wellness programs do? Evidence from the Illinois workplace wellness study. *The Quarterly Journal of Economics*, *134*(4), 1747–1791.

Karakolis, T. & Callaghan, J. P. (2014). The impact of sit–stand office workstations on worker discomfort and productivity: A review. *Applied Ergonomics*, *45*(3), 799–806.

Kroese, F. M., Marchiori, D. R. & de Ridder, D. T. D. (2016). Nudging healthy food choices: A field experiment at the train station. *Journal of Public Health*, *38*(2), e133–e137.

Kwak, L., Kremers, S. P. J., Van Baak, M. A. & Brug, J. (2007). A poster-based intervention to promote stair use in blue- and white-collar worksites. *Preventive Medicine*, *45*(2/3), 177–181.

Lewis, A. (2012). *Why Nobody Believes the Numbers: Distinguishing Fact from Fiction in Population Health Management*. John Wiley and Sons.

Lewis, A. & Eves, F. (2012). Prompt before the choice is made: Effects of a stair-climbing intervention in university buildings. *British Journal of Health Psychology*, *17*(3), 631–643.

Meyer, P., Kayser, B., Kossovsky, M. P., Sigaud, P., Carballo, D., Keller, P. F., Eric Martin, X., Farpour-Lambert, N., Pichard, C. & Mach, F. (2010). Stairs instead of elevators at workplace: Cardioprotective effects of a pragmatic intervention. *European Journal of Cardiovascular Prevention and Rehabilitation*, *17*(5), 569–575.

Michie, S., van Stralen, M. M. & West, R. (2011). The behaviour change wheel: A new method for characterising and designing behaviour change interventions. *Implementation Science, 6*(42).

Miller, R., Williams, P. & O'Neill, M. (2018). *The Healthy Workplace Nudge: How Healthy People, Culture, and Buildings Lead to High Performance.* John Wiley and Sons.

Moloughney, B. W., Bursey, G. E., Fortin, R. B., Morais, M. G. & Dang, K. T. (2019). A Multicomponent intervention to encourage stair use in municipal buildings. *American Journal of Health Promotion, 33*(1), 57–69.

Neuhaus, M., Eakin, E. G., Straker, L., Owen, N., Dunstan, D. W., Reid, N. & Healy, G. N. (2014). Reducing occupational sedentary time: A systematic review and meta-analysis of evidence on activity-permissive workstations. *Obesity Reviews, 15*(10), 822–838.

Nicoll, G. & Zimring, C. (2009). Effect of innovative building design on physical activity. *Journal of Public Health Policy, 30*(Suppl. 1), S111–S123.

Nocon, M., Müller-Riemenschneider, F., Nitzschke, K. & Willich, S. N. (2010). Increasing physical activity with point-of-choice prompts – A systematic review. *Scandinavian Journal of Public Health, 38*(6), 633–638.

Olsson, T., Jarusriboonchai, P., Woźniak, P., Paasovaara, S., Väänänen, K. & Lucero, A. (2020). Technologies for enhancing collocated social interaction: Review of design solutions and approaches. *Computer Supported Cooperative Work, 29*(1), 29–83.

Sailer, K. & Koutsolampros, P. (2021). Space syntax theory: Understanding human movement, co-presence and encounters in relation to the spatial structure of workplaces. In R. Appel-Meulenbroek & V. Danivska (Eds.), *A Handbook of Theories on Designing Alignment Between People and the Office Environment* (pp. 248–260). Routledge.

Sawyer, A., Smith, L., Ucci, M., Jones, R., Marmot, A. & Fisher, A. (2017). Perceived office environments and occupational physical activity in office-based workers. *Occupational Medicine, 67*, 260–267.

Segerstrom, S., & Sephton, S. 2010. "Optimistic expectancies and cell-mediated immunity: The role of positive affect." *Psychological Science, 21*(3), 448–455.

Service, O., Hallsworth, M., Halpern, D., Algate, F., Gallagher, R., Nguyen, S., Ruda, S. & Sanders, M. (2015). *EAST: Four Simple Ways to Apply Behavioural Insights.* Retrieved from: https://www.bi.team/wp-content/uploads/2015/07/BIT-Publication-EAST_FA_WEB.pdf

Spreitzer, G., Bacevice, P. & Garrett, L. (2020). Workplace design, the physical environment, and human thriving at work. In O. B. Ayoko & N. M. Ashkanasy (Eds.), *Organizational Behaviour and the Physical Environment* (pp. 235–250). Routledge.

Sternberg, E. 2009. *Healing Spaces: The Science of Place and Well-Being.* Harvard University Press: Cambridge, MA.

Swenson, T. & Siegel, M. (2013). Increasing stair use in an office worksite through an interactive environmental intervention. *American Journal of Health Promotion, 27*(5), 323–329.

Thaler, R. H. & Sunstein, C. R. (2008). *Nudge: Improving Decisions about Health, Wealth, and Happiness.* Yale University Press.

Torbeyns, T., de Geus, B., Bailey, S., De Pauw, K., Decroix, L., Van Cutsem, J. & Meeusen, R. (2016). Bike desks in the office. *Journal of Occupational and Environmental Medicine, 58*(12), 1257–1263.

Van Woerkom, M. (2021). Building positive organizations: A typology of positive psychology interventions. *Frontiers in Psychology, 12*, 769782.

Venema, T. A. G. & Jensen, N. H. (2023). We meat again: A field study on the moderating role of location-specific consumer preferences in nudging vegetarian options. *Psychology and Health, 39*(10), 1337–1351.

Venema, T. A. G., Kroese, F. M. & De Ridder, D. T. D. (2018). I'm still standing: A longitudinal study on the effect of a default nudge. *Psychology and Health, 33*(5), 669–681.

Venema, T. & van Gestel, L. (2021). Nudging in the workplace: Facilitating desirable behaviour by changing the environment. In R. Appel-Meulenbroek & V. Danivska (Eds.), *A Handbook of Theories on Designing Alignment between People and the Office Environment* (1st Edition). Routledge, pp. 222–235.

Wilkerson, A. H., Usdan, S. L., Knowlden, A. P., Leeper, J. L., Birch, D. A. & Hibberd, E. E. (2018). Ecological influences on employees' workplace sedentary behavior: A cross-sectional study. *American Journal of Health Promotion, 32*(8), 1688–1696.

World Health Organization (2006). *Constitution of the World Health Organization.* Retrieved from: https://www.who.int/publications/m/item/constitution-of-the-world-health-organization

Zhu, X., Yoshikawa, A., Qiu, L., Lu, Z., Lee, C. & Ory, M. (2020). Healthy workplaces, active employees: A systematic literature review on impacts of workplace environments on employees' physical activity and sedentary behavior. *Building and Environment, 168,* 106455.

9 Organisational and national cultures

Editors' introduction

Cultures are the glue that ties together the people working at an organisation (organisational culture) or all of the people living in a politically defined zone (national culture). They define "how things get done around here" whether that here is Ford or Romania.

As more cultures mix in a single workplace and as people design for other cultures, understanding what drives members of cultures to professional success, and how they even define "success," is becoming particularly important. Growing up in our own culture we know its core drivers, how it "works," either consciously or unconsciously. Nonverbal communication, silent signalling via design elements, is particularly important for sending messages at a culture-wide level, but only members of a particular culture will ever really know what is being "heard" when those missives are transmitted – another reason why research with probable users is important.

Design can support cultures, as discussed in the pages that follow, allowing the people living in them to reach their full potential. What design can't do is magically and instantaneously change cultures. Cultures evolve over a long term, years, and can't be switched on or off. A culture in place today will influence work-related expectations and experiences over the long term. Any new workplace that springs up will be assessed in light of previously existing cultures and used accordingly and with existing-culture consistent outcomes.

DOI: 10.1201/9781003390848-9

Researcher perspective

Sally Augustin

Overview

When the design of workplaces and the objects in them aligns with the national and organisational cultures of users, wellbeing gets a boost and professional performance is also elevated.

National culture-aware workplace design

Understanding how national culture should influence workplace design is becoming more important as people from different cultures are using the same workplaces, as organisation offices become more and more geographically distributed, and as people in one part of the world regularly find themselves developing offices, whereas people from another area on the planet will find themselves working.

Designing to support national culture matters. For instance, Grenness (2015) reports that the office design which Telenor (a Norwegian company) implemented in its home office worked well there but not in Asia where national culture was very different.

Hofstede, Hofstede, and Minkov (2010) define national culture as "the unwritten rules of the social game. It is the collective programming of the mind that distinguishes the members of one group or category of people from others." The empirically derived system that this Hofstede-led team presents is straightforward and has clear-cut implications for workplace design. The group identifies 6 major criteria for distinguishing cultures and the high-level consequences of high or low scores on each factor.

* *Individualistic-Collectivistic* – People in more individualistic cultures are more likely to perceive that they are independent of others and the reverse is true in collectivist cultures where people feel more bonded to groups and the desire to conform to the "design rules" for their areas (such as using or not using particular colours) can be strong. Those who hail from more individualistic countries have higher expectations of being able to have privacy when they choose than people from more collectivistic ones while those who are from more collectivistic cultures are more amenable to sharing resources, such as printers or bathrooms that people from more individualistic ones. Also, people from more individualistic cultures are more likely to try to make changes to the environments in which they find themselves, by moving furniture or screens between desks, for instance, than people from more collectivistic cultures ones and that needs to be taken into account during the design process, with some flexibility

built into use, or a space will "ugly up" fast. Hofstede, Hofstede, and Minkov (2010) report that countries relatively high on individualism include the United States, Great Britain, Sweden, and Germany while countries high on collectivism include Venezuela, South Korea, Singapore, and China. Information on scores on this and other cultural factors are available in the 2010 resource and also at the Hofstede team's authorised websites.

- *Tolerance of an uneven distribution of power* – Countries that score higher on this factor are said, in the Hofstede team's terminology to have greater power distance. The consequences of a higher score for workplace design would be being more tolerant/accepting of people with more power having access to more workplace amenities such as an executive-only fitness centre, for instance. Some countries with relatively high-power distance according to Hofstede's team are Russia, Indonesia, and China while countries with relatively low power distance are Israel, New Zealand, Sweden, Great Britain, and the United States.
- *Toughness* – Hofstede and colleagues officially label this factor "masculine-feminine," with masculine countries being tougher. The team shares that in more feminine countries, quality of life is more important and achievement is less of a motivator than it is in masculine ones. Behaving in an environmentally responsible way is even more important in feminine countries than in masculine countries. Relatively more masculine countries, according to the Hofstede team, are Japan, Italy, Germany, Great Britain, China, and the United States while countries to the opposite end of the spectrum are Sweden, the Netherlands, and Costa Rica.
- *Tolerance for uncertainty* – Countries that are towards the high end with tolerance of uncertainty are more comfortable with ambiguity and see less need for rules, compared to people who are less accepting of uncertainty—which has repercussions for things such as systems for programming sites and how stringently they are enforced. People who are more accepting of uncertainty are more attracted to novel design solutions and less concerned about cleanliness than people who are less tolerant. The orientation to cleanliness can have implications for material selections as well as for designing for ease of cleaning (e.g., vacuuming). Countries with higher scores on the uncertainty avoidance scale are Greece, Russia, Turkey, and Japan. Countries with relatively lower scores on this factor, according to the Hofstede-led team, are Sweden, China, Great Britain, and the United States.
- *Time orientation* – Some countries have a longer term orientation, but others are the reverse. In areas with a long-term orientation, there is more concern about the future and achieving set goals and in countries with a short-term orientation, there is more interest in trends as well as traditions. People in countries with a future orientation can be particularly attuned to the return on investment of projects undertaken. Nations with a relatively long-term orientation are China, Japan, Germany, the Netherlands, and Russia while those with a short-term orientation include Nigeria, Venezuela, and Egypt.
- *Indulgent-Restrained* – Hofstede and associates have added this factor relatively recently and its consequences for workplace design are not yet established. People in indulgent cultures feel freer to enjoy life than those in more retrained

ones, for example. More indulgent cultures are found in Mexico, Sweden, the United States, and Australia while more retrained ones are present in Hong Kong, Romania, and Russia.

Jordan (2000) reviewed a number of studies to develop guidelines for successful design strategies in different sorts of cultures. He reports that

links were made between the five cultural dimensions and people's preferences and tastes with respect to what a product design should communicate through its aesthetics. Some of these links are as follows.

- *Power distance* – High status [for relatively high scoring cultures, where more uneven distributions of power are more widely accepted], youthfulness [in countries where this sort of uneven distribution is not as generally accepted].
- *Individuality* – Expressiveness (relatively high scorers on individuality), familiarity (relatively more collectivist countries).
- *Toughness [masculine-feminine]* – performance (in relatively high scoring, more masculine cultures), artistry (relatively more feminine ones).
- *Uncertainty avoidance* – Reliability (relatively high uncertainty avoiding groups), novelty (relatively low scorers).
- *Long-term orientation* – Timelessness (in cultures where long-term orientation is stronger), fashionableness (in countries where short-term orientation prevails).

Studies done using Hofstede's system have also established the following.

- Nonverbal communication is particularly important in collectivistic countries (de Mooij and Hofstede, 2011).
- Cai and Zimring (2019) report that for collectivistic cultures there are

    greater needs for group collocation, physical proximity, and better visual connections that can bring group members together and create internal solidarity. A space with higher visual connectivity and accessibility is helpful to create more opportunities for casual face-to-face interactions and maintain the bond among the group members. In addition, the focus on the distinction between in-group and out-group requires identifiable boundaries between different user groups.

- People living in countries with a long-term orientation are not as willing to pay for convenience as those from short-term ones.

(de Mooij and Hofstede, 2002)

Although this classification system was originally developed in the 1970s, more recently, collected data indicates that it is still valid, with specific countries having profiles today consistent with those present earlier (de Mooij and Hofstede, 2010).

When considering Hofstede, Hofstede, and Minkov's system, it's important to bear in mind that their countrywide categorisations are just that, countrywide, and any individual person living in a country might feel/think differently than other people who live in the same country.

A number of other factors, such as responses to particular colours, can also be linked to national culture as can ways of absorbing information, for example, how eyes move when art is being reviewed and what art is preferred, and how much eye contact is desirable. Readers are encouraged to learn more about these differences using tools such as scholar.google.com.

Organisational culture aligned office design

Organisational culture is much like national culture but (except for the very largest of groups) at a smaller scale. Both function as generally unspoken "playbooks," in use over extended periods of time, that guide thoughts and actions and help newcomers understand "how things get done around here." One culture can be found across an entire firm, etc., or different groups within an entity may vary in culture.

Workplace design that's consistent with organisational culture has been linked to greater wellbeing and cognitive performance for users (Peponis et al., 2007; Shein, 1990; West and Wind, 2007). Nanayakkara and colleagues (2021) share that "the critical achievement of workspace design is to integrate the cultures, values and behaviours of organisations to meet their ultimate goals."

Space users can effectively read messages related to culture sent by the environments they use (Becker and Steele, 1995; Schein, 1990), and the only way to be sure what messages they are pulling from a space is to ask them. Schein (1990) states that "it is desirable to distinguish three fundamental levels at which culture manifests itself: (a) observable artifacts, (b) values, and (c) basic underlying assumptions. When one enters an organisation one observes and feels its artifacts."

Research indicates that users believe the messages sent by the physical forms of workplaces more reliably present an organisation's actual culture than mission and value statements (Becker and Steele, 1995).

One of the systems for classifying organisational cultures that's most useful to designers was developed by Cameron and Quinn (for example, 2006). Their diagnostic survey is the OCAI (Organizational Culture Assessment Instrument) and it is widely available online.

Cameron and Quinn categorise organisational cultures as follows.

- *Hierarchical*, with many formal structures/policies in place, "The long-term concerns of the organisation are stability, predictability, and efficiency."
- *Market*, where business results such as profitability and market focus attention, "The core values that dominate market-type organisations are competitiveness and productivity."
- *Clan*, groups at which teamwork and employee wellbeing are given a great deal of attention,

the organization is in the business of developing a humane work environment, and the major task of management is to empower employees and facilitate their participation, commitment, and loyalty ... Success is defined in terms of internal climate and concern for people.

- *Adhocratic*, organisations at which "innovative and pioneering initiatives are what leads to success, that organizations are mainly in the business of developing new products and services and preparing for the future ... they can reconfigure themselves rapidly when new circumstances arise."

Cameron and Quinn (2006) also use a single adjective to describe each culture.

- *Hierarchy* – Controlling.
- *Market* – Competitive.
- *Clan* – Collaborative.
- *Adhocracy* – Creative.

The design implications of Cameron and Quinn's model are straightforward. For example, establishing workplace design protocols (based on whatever parameters are important to the organisation, be they job function, rank, or something else), and adhering to them, is very important in hierarchical cultures. Spaces used by clan cultures will be of the highest design standards affordable, tuned to support employee wellbeing, and group areas where clan-notes can socialise are important spaces for this culture. People working in market cultures are driven to "win"; however, winning is defined in their group, which might deal in commodities or transport payloads to outer space, for example, and spaces will be designed to support doing just that, however, that can be achieved — via giant boards visible to all that present information useful at a particular moment (such as market prices for something at that instant) or very open spaces so workers can hear trades being executed by others, or something else entirely. In adhocratic cultures, creativity and innovation are paramount, and spaces must support just that, as outlined in Chapter 15 of this text.

Conclusion

Although both are clearly important, national culture more significantly influences optimal workplace design than organisational culture. As Hofstede, Hofstede, and Minkov (2010) report "Nationality defines organizational reality ... [research presented] demonstrated six ways in which national cultures differ; all of these have implications for organization and management processes."

When workplace design recognises, respects, and reflects the national and organisational cultures of users, the wellbeing and performance of those users and the organisations that employ them are elevated.

Practitioner perspective

Arnold Levin

The culture conundrum

Within the design profession, there is often ambiguity concerning how one defines culture. When asking organisations to define their culture, we hear words such as collaborative, social, and collegial used as adjectives. Designers use the same words to describe workplace solutions when attempting to connect their design solutions to a client's culture. The problem with this thinking and approach is that first we take for granted the efficacy of the words and we assume because one says it is so, therefore it is. Design responses look to imagery and space typologies to reinforce the words. Furniture benching as a typology has been used to symbolise collaboration. A coffee bar symbolises collaboration as well as social interaction. The real question is: do these typologies and images make an organisation collaborative or social?

This conundrum highlights a flaw in how we have been viewing and assessing culture. We have been focusing on "workplace" culture rather than "organisational" culture. It's much easier to miss the big picture when all we are focusing on is workplace culture. By its nature, the relationship between the term workplace and its physical manifestation conjures imagery while ignoring the fact that the physical workplace is a reflection of the organisations' culture and that the work environment could be a reflection of that organisation. What I will explore in this chapter is the role of organisational as well as national culture in informing the design of the office and how they impact both performance outcomes and defining how "things get done" within that particular organisation.

Defining culture

We most commonly think of defining culture as an organisation's values and beliefs, what often becomes the norm within the working environment. The organisational designer, Naomi Stanford writing in her book *Organizational Culture* (Stanford, 2010), highlights five varying perspectives of organisational culture thereby illustrating the ambiguity in understating just what culture is within an organisation and how to focus in on it in order to impact performance especially when it comes to the design of the office. Among the five definitions cited, the one that comes closest to the meaning used most commonly within the workplace design community is:

> A system of shared values, defining what is important, and norms, defining appropriate attitudes and behaviors, that guide members' attitudes and behaviors.
>
> (Stanford, 2010)

The other perspective she cites is:

> The pattern of values, norms beliefs, attitudes and assumptions that may not have been articulated but shape the ways in which people behave and things get done. Values refer to what is believed to be important about how people and the organisations behave. Norms are the unwritten rules of behavior.
>
> (Stanford, 2010)

Reinforcing one of the overarching sub-texts of this text, she adds:

> Beyond these five definitions is the popular one of uncertain origin defining culture as 'the way we do things round here.
>
> (Stanford, 2010)

Work culture

As it relates to work, understating an organisation's culture needs to move beyond the adjectives and descriptions, and into a broader view of how that organisational culture influences how work actually gets done. Referring again to Stanford's definition of culture and how it is viewed within the context of the broader organisation, she offers ten organisational characteristics (Stanford, 2010):

- a story or stores,
- a purpose or set of values,
- a strategy,
- an attitude to people,
- a global mindset,
- a relationship networks,
- a digital presence,
- a reputation,
- a customer proposition,
- a horizon scanning ability.

National culture

When it comes to national culture and its role in workplace design solutions, there has been little research and applied solutions that speak to this facet of culture. Taken in combination, workplace culture and national culture remain an enigma. One of the best perspectives on the role of national culture and how it informs workplace performance comes from Juriaan van Meel, a Dutch workplace strategist who turned his PhD dissertation on office design and national context into the book: *The European Office* (van Meel, 2000). As part of his study, he examined five European countries and their approach to workplace design and strategy (The UK, Germany, Sweden, Italy, and The Netherlands). The study focused on five frameworks of analysis: Urban Setting, Market Conditions, Labour Relations, Regulations, and Culture.

One aspect of van Meel's thesis that puts the importance of including an acknowledgement of national culture as part of workplace strategy recommendations is first recognising the reality of significant differences between national cultures of countries and how that can inform perceptions around strategies that may be erroneously considered universal. His study by its structure points to the intricate differences between the European cities on the influence of different national cultures on the perception of workplace utilisation as well as performance. He also points to the fact that within Europe, national context has a larger impact on office design than in the United States (van Meel, 2000):

There may be a single market, but there is no single business culture ...

Regarding national cultural differences, he notes the differences between acceptance and adoption of theoretical strategies between the United States and Europe:

One reason why in the 1920's, Taylors ideas about scientific office management were less easily adopted in Europe that the US was European business culture still seemed rooted in the high-status high trust of the male clerks of the 19th century.

(van Meel, 2000)

Conversely, he notes:

Cultural differences are also an explanation for different responses in the 1960's to the office landscape strategies.

[developed by the German Quickborner Team]

The role of culture in office design

The interpretation of culture within office design has been viewed in large part through imagery both tacitly and explicitly. However, these images often took on different meanings over decades. Take for instance, the imagery conjured up by the Taylorist office of the 1920s. As it was originally conceived, the rows of identical desks lined up were meant to implement a vision of scientific management and the assembly line, from which scientific management as a theory was defined. Seventy years later, the contemporary vision of what was planned as benching was then imagined as a symbol of openness and collaboration, and not the regimentation of the assembly line.

The role of imagery also had explicit and tacit expressions. While the use of open spaces has been used to signify transparency and collaboration, office design does not always explicitly state its intended purpose to reflect the culture of the organisation. Creating spaces through extensive use of private offices and cubicles may not be by design a reflection of that organisation's culture, but the very nature of the so-called "cube farms" creates an image of a culture of impersonalisation and hierarchy. Writing in *The Changing Workplace*, author Frank Duffy cites:

> The workplace is a statement of beliefs, even when that function is not the stated objective.
>
> (Duffy, 1992)

Within the world of office design, imagery as an expression of culture has played a significant role. While imagery is important to physically manifest the messaging of an organisations culture, it has also sometimes served to mask the realities of an organisations culture. It can create meaning where there is no substance, or it misses the opportunity to create more substantive design solutions that provide more meaningful value to the organisation. This is not to minimise the role of image as an important means of portraying an organisations culture, but to create an awareness that by relying solely on one aspect of translating an organisations culture into the workplace, we are missing opportunities.

In fact, Weick in *Sensemaking in Organisations*, points to what he refers to as "construction of identity" (Weick, 1995). Identities are formed within organisational settings that form perceptions. However, by focusing only on image as solutions and representations of workplace, design problems are often limiting, sometimes misrepresentational and limit the strategist or designer in their analysis of culture, thereby falling short on the potential promise of the workplace to connect design solutions to the organisations business. The source of this mindset can be traced back to the schism between the early architects of the modern movement and the rise and role of corporate architecture from the post-World War II period. Peter Blake in his book *No Place Like Utopia* (Blake, 1993) traces that split in approach. Discussing the vision held by pre- and post-war architects such as Bruer and others, he notes that their vision of the function of the modern movement was to first and foremost solve society's problems, primarily housing. Problem-solving was their ethos and reason for practice. With the post-World War II rise of corporate headquarters emanating in the United States, architects focused on brand and imagery of their corporate clients as architectural solutions. The British architectural philosopher, Neal Leach takes this point further in his book *The Anaesthetics of Architecture* (Leach, 1999). His thesis is that architects had become so preoccupied with imagery and form, from modernism to post-modernism and every design movement trend following, that they became desensitised to the true power and purpose of design as a problem-solving endeavour.

Aligning culture to design solutions: Moving beyond image and language

One way of better aligning work culture to design solutions is to use a different lens through which to analyse and understand an organisation's culture. The typical means of currently identifying culture within an organisation comes through visioning sessions or workshops where the strategist or designers probe the client for their vision or definition of their culture. Employee surveys as well as observational analysis or ethnographic research may be added to the methodology.

While these research venues should be part of the "toolbox" of cultural analysis, for the most part, it omits examining the organisation from a broader and more meaningful perspective. It also results in solutions that primarily involve culture as image as I have suggested above. Work culture needs to be aligned with the culture of the organisation, work being a manifestation or a product of that organisation. Hence, the need to look at culture in a more comprehensive manner. Analysing a work culture to create better performance needs to be linked back to the design of that organisation's business model in order to truly result in better performance but also as a means to measure that performance. Joanne Martin, a professor of organisational behaviour at Stanford University suggests going beyond the surface understanding of culture and seeking "an in depth understanding of the patterns of meanings that link these manifestations together …" (Martin, 2002), thereby underscoring the importance of going beyond the imagery of culture.

A more meaningful lens through which to understand a client's culture would be through organisational design, or the components and elements that comprise the make-up of an enterprise. This perspective allows one to take the surface meaning of culture as comprising of those shared values and norms to better viewing these values and norms from how they are actually made real. The current gap in understanding culture and interpreting it through design solutions is not from the definition, it is from understating how the organisation made these norms and values real. The only way to do that is through the lens of organisational design. Organisational design is simply the design of organisations which comprises the various elements that make up that enterprise.

This framework also allows us to understand and approach an organisations culture in one of two ways. For some workplace solutions, one might be trying to enable an existing culture through reinforcing those values and meanings in the physical workspace. In other instances, the goal could be to use the workplace as a tool to facilitate a desired change in a culture. In order to tackle either of these goals, we have to change the perspective of what we are actually analysing. It is here that using organisational design frameworks provides that knowledge and focus that currently does not reside in the traditional toolbox of the workplace design profession. The ability of organisational design as a framework goes to the heart of what organisational design is all about and the possibilities it affords us in better tackling the culture conundrum through connecting the explicit qualities of the organisations to the physical. The organisational designer, Jeroen van Bree, reports that:

> One of the aspects that set organization design apart is its focus on formalizable elements such as structure, processes, and roles and a claim that we can design or redesign those elements to achieve intended outcomes…and its claim that we can make or remake these artifacts to change existing outcomes into preferred ones.
>
> (van Bree, 2021)

There are numerous frameworks that organisational designers use to assess an organisations' design and offer solutions for organisational improvement. One framework that I have made use of is Jay Galbraith's (1995) Star Model, often referred to as a congruence model (Figure 9.1). In my MBA dissertation, *Workplace Design, A Component of Organizational Strategy* (Arnold, 2015), I demonstrated, first, the relationship between workplace design and organisational design, both in terminologies, processes, and interdependence. I went on to show how those analytical tools employed by organisational designers could be used by workplace strategists and designers in their assessment process.

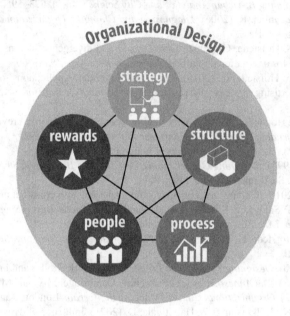

Figure 9.1 Galbraith's Star Model of organisations (based on Galbraith, 1995, adapted from Wikimedia Commons).

The *Star Model* uses as the source of analysis five basic organisational components: strategy, structure, processes, people, and rewards (the organisations' strategy, how they are organised or structured, i.e. formal vs. informal, hierarchical vs. networked; their work processes; the people who make up that organisation; and the reward systems that serve to reinforce desired behaviours and outcomes). For any business to be successful, each of these components needs to be aligned or congruent. It is these five organisational components that create and form the culture (or norms and values) of the organisation.

The role of culture, how it is identified and analysed, is even more critical in this so-called post-pandemic world of work, where the boundaries between where we work and how we work are being blurred to an extent never observed before. The need for a more rigorous approach to understanding one's work culture is more dependent now on having the ability and tools to go beyond imagery and create more substantive connections and solutions involving culture.

References

Arnold, L. (2015). *Workplace Design: A Component of Organizational Strategy, MBA Dissertation.* Harrow Business School, University of Westminster.

Becker, F. & Steele, F. (1995). *Workplace by Design: Mapping the High-Performance Workscape.* Jossey-Bass: San Francisco, CA.

Blake, P. (1993). *No Place Like Utopia, Modern Architecture and the Company We Kept.* New York: Alfred A. Knopf.

Cai, H. & Zimring, C. (2019). Cultural impacts on nursing unit design: A comparative study on Chinese nursing unit typologies and their us counterparts using space syntax. *Environment and Planning B: Urban Analytics and City Science, 46*(3), 573–595.

Cameron, K. & Quinn, R. (2006). *Diagnosing and Changing Organizational Culture.* San Francisco: Jossey-Bass.

de Mooij, M. & Hofstede, G. (2002). Convergence and divergence in consumer behavior: Implications for international retailing. *Journal of Retailing, 78,* 61–69.

de Mooij, M. & Hofstede, G. (2010). The Hofstede model: Applications to global branding and advertising strategy and research. *International Journal of Advertising, 29*(1), 85–110.

de Mooij, M. & Hofstede, G. (2011). Cross-cultural consumer behavior: A review of research findings. *Journal of International Consumer Marketing, 23,* 181–192.

Duffy, F. (1992). *The Changing Workplace.* London: Phaidon Press.

Galbraith, J. (1995). *Designing Organizations, An Executive Briefing on Strategy, Structure and Process.* San Francisco, CA: Josey Bass.

Grenness, T. (2015). Culture matters: Space and leadership in a cross-cultural perspective. In A. Ropo, P. Salovaara, E. Sauer & D. de Paoli (Eds.), *Leadership in Spaces and Places.* Northampton, MA: Edward Elgar Publishing, 199–214.

Hofstede, G., Hofstede, G. J. & Minkov, M. (2010). *Cultures and Organizations.* New York: McGraw Hill.

Jordan, P. (2000). *Designing Pleasurable Products.* New York: Taylor and Francis.

Leach, N. (1999). *The Anaesthetics of Architecture.* Cambridge, MA: The MIT Press.

Martin, J. (2002). *Organizational Culture: Mapping the Terrain.* London: Sage Publications.

Nanayakkara, K., Wilkinson, S. & Halvitigala, D. (2021). Influence of dynamic changes of workplace on organisational culture. *Journal of Management and Organization, 27*(6), 1003–1020.

Peponis, J., Bafna, S., Bajaj, R., Bromberg, J., Congdon, C., Rashid, M., Warmels, S., Zhang, Y. & Zimring, C. (2007). Designing space to support knowledge work. *Environment and Behavior, 39*(6), 815–840.

Schein, E. (1990). Organizational culture. *American Psychologist, 45*(2), 109–119.

Stanford, N. (2010). *Organizational Culture, Getting it Right.* The Economist Newspaper.

van Bree, J. (2021). *Organization Design.* Utrecht: Berenschot.

van Meel, J. (2000). *The European Office: Office Design and National Context.* Rotterdam: 010 Publishers.

Weick, K. E. (1995). *Sensemaking in Organizations.* London: Sage Publications.

West, A. and Wind, Y. (2007). "Putting the organization on wheels: workplace design at SEI." *California Management Review, 49*(2), 138–153.

10 Occupant personality

Editors' introduction

Oseland (2022) reminds us that "Personality is an individual's unique set of traits and relatively consistent pattern of thinking and behaviour that persists over time and across situations." The occupant may adjust or adapt their personality to fit in with the organisational culture and to cope with different situations, but their underlying base personality is quite fixed.

Tailoring a home to support a single user with a single personality profile is easy, but recognising and respecting personality differences in a large corporate workplace is more of a challenge. However, it is made slightly easier because people with similar personalities are attracted to similar jobs, that they do well (Holland, 1966). That people with similar personalities tend to be successful in particular roles long term does not guarantee that everyone in a particular role will have the same personality profile, but it does make likely that enough do so that a workspace can be designed to support occupants' most probable personality type.

Designing workplaces to support a range of personality types, rather than say just extroverts, is worth the effort. Susan Cain (2012) explains that doing so increases the likelihood that people will perform better at work. Different personality types thrive in different environmental conditions, their stress levels, wellbeing are affected as well as their performance.

DOI: 10.1201/9781003390848-10

Researcher perspective

Nigel Oseland

Preamble

Like all undergraduate psychologists, I was taught personality theory and how it affects perception, reactions and behaviour. As a postgraduate in environmental psychology and a workplace consultant, my current interest is how personality impacts an individual's workplace requirements and how, in turn, they affect comfort, wellbeing and performance at work.

Origins of personality theory

The word "personality" is derived from *persona* which is Latin for "mask" – the mask we present to the world. Interestingly, there is no consensus among psychologists on a single definition of personality, but my following definition (Oseland, 2022) consolidates the common viewpoints.

> *Personality is an individual's unique set of traits and relatively consistent pattern of thinking and behaviour that persists over time and across situations.*

Many factors influence personality including heredity, culture, experiences, family and people that we engage with. Consequently, personality is stable but not necessarily fixed, it is a bias towards core traits or characteristics that steer our responses and behaviours. So, we are likely to prefer environments that support our inherent behaviour and underlying personality.

The earliest records of a structured theory of personality date back many centuries. Greek physicians Hippocrates (circa 400 BC) and Galen (circa 150 AD) proposed four different personality types referred to as humours or temperaments: phlegmatic (calm), sanguine (optimistic), melancholic (depressed) and choleric (irritable). They proposed that the temperaments were caused by chemical imbalances in the body. While bizarre, neuropsychologists acknowledge that certain chemicals in the brain affect mood and behaviour.

Carl Jung is credited with first categorising people as introverts and extroverts. The extroversion scale has influenced most subsequent theories of personality. Jung also suggested there are four basic ways of reacting (termed functions) which combined with introversion or extroversion form eight different personality types. The sensing (S) and intuition (N) functions relate to the way individuals acquire information whereas the thinking (T) and feeling (F) functions relate to reaching decisions. (Figure 10.1).

Another approach to understanding personality is to identify and describe it in terms of traits, or characteristics. There are so many descriptors of personality that it is difficult to make sense of them. Allport and Odbert (1936) initially found some 17,953 descriptors but reduced their list to 4,504. Later Cattell (1947) reduced the descriptors further to 171 and ultimately created his *Sixteen Personality Factors* (16PF) model.

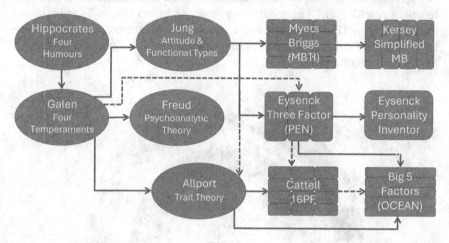

Figure 10.1 The development of modern personality theories (Oseland, 2022).

Eysenck and Eysenck's (1975) two super-traits model was derived directly from Jung's theories and even refers to the four temperaments. Cattell's 16PF model was considered to have too many superfluous dimensions, so two personality dimensions were proposed: extroversion (E) and neuroticism (N) (Figure 10.2).

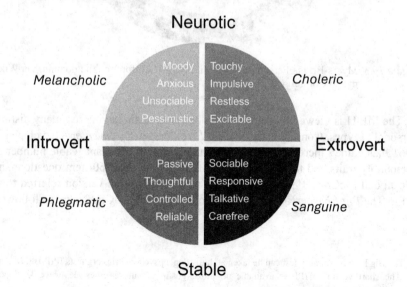

Figure 10.2 Super-trait model of personality (adapted from Eysenck and Eysenck, 1975).

Practical personality models

Briggs and Myers (1987) elaborated on Jung's personality theory by adding a function which indicates the way people interact with the environment. Judgers (J) prefer an organised, stable environment, whereas perceivers (P) are flexible and spontaneous. Adding these dimensions to those of Jung creates a matrix of functions resulting in 16 personality types, or the Myers Briggs Type Indicators (MBTIs). The MBTI is popular in modern business management and used for evaluating and developing teams and determines the suitability of potential employees for job roles (Figure 10.3).

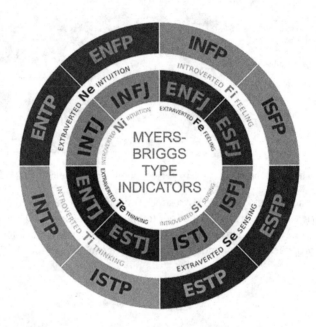

Figure 10.3 Myers-Briggs Personality Type Indicator (Offnfopt, public domain, via Wikimedia Commons).

The MBTI is viewed by some psychologists as again having too many distinct personality types. Consequently, the *Big Five Personality Inventory*, developed in 1961, has gained increasing popularity largely due to the manageable number of personality traits and the practicality of its relatively short 60-item questionnaire (Costa and McCrae, 1992),[1] and there is a shorter ten-item version referred to as TIPI.[2] The *Big Five* personality traits, also referred to as OCEAN, are as follows.

---

[1]   The Big Five Personality test can be taken at: https://openpsychometrics.org/tests/IPIP-BFFM/.
[2]   The short version, TIPI, is available at: https://gosling.psy.utexas.edu/scales-weve-developed/ten-item-personality-measure-tipi/.

- *Openness (to new experiences)* – creative/curious versus conventional/conservative.
- *Conscientiousness* – reliable/organised versus casual/unfocussed).
- *Extroversion* – gregarious/sociable versus reserved/solitary.
- *Agreeableness* – cooperative/empathetic versus challenging/antagonistic).
- *Neuroticism* – anxious/nervous versus calm/confident).

It may be considered impractical to design a workplace for a variety of personality types. Fortunately, different job functions attract similar personality types (Nguyen, 2020; Schaubhut and Thompson, 2008). Apparently, sales and marketing will attract more extroverts who seek stimulation and take more risks, focus more on the big concept than details and thrive on meeting and socialising with people. In contrast, introverts are more likely to prefer research and analysis, and a finance team involved in heavy processing of data will usually attract a high proportion of introverts who are better suited for conducting detailed tasks. Therefore, the workplace for a business unit could be designed to suit the more common personality types in the team.

The MBTI and the *Big Five* both include an introversion-extroversion (I/E) scale. Of all the personality types, I have found the I/E scale the most useful in researching and advising on work environments. This is partly due to the logic and ease of explanation, and because it is more extensively used in research studies. For example, the impact of noise, a stressor, on the performance of introverts and extroverts has been well researched, notably by Donald Broadbent and colleagues at Cambridge University in the 1950s (Broadbent, 1958). The research shows that in general, extroverts cope, and even perform, better in stimulating environments whereas introverts find such environments over-stimulating and distracting which in turn hinders their performance.

One explanation for the difference in response between introverts and extroverts is our innate level of arousal. Yerkes and Dodson (1908) proposed an inverted U-shape relationship between performance and level of arousal. People perform better if they are stimulated or motivated, but there is a limit as too much stimulation leads to stress and reduced performance. In contrast, under-stimulation leads to lethargy thus reducing performance. It is proposed that extroverts have a low natural level of arousal and so seek stimulation whereas introverts have a higher level of arousal and so prefer low levels of stimulation.

Acoustics, noise and personality

I have spent many years researching psychoacoustics and how to reduce noise distraction in offices. In a series of online surveys with 2,145 respondents, across a range of organisations and countries, Paige Hodsman and I found that 50% of workers consider workplace noise to adversely affect their wellbeing and increase stress, and 67% rated the effect of noise on performance as negative (Oseland and Hodsman, 2017). Our survey also found that introverts were more negatively affected by noise. In addition, we also found that the more neurotic (nervous, apprehensive), less agreeable (less empathetic towards colleagues) and, to some extent, more conscientious (diligent) are also more adversely affected by noise in open plan offices.

Susan Cain (2012) reminded us that introverts are often overlooked despite making considerable contributions to society and that the workplace contains similar proportions of introverts and extroverts. Nonetheless, the modern workplace is often designed solely with the extrovert in mind with a tendency to create open plan, noisy, buzzy and crowded environments that are stimulating to facilitate interaction. Yet, these environments can cause distraction, higher stress and poor performance for introverts, especially those involved in complex and detailed analysis, i.e. roles that are often attractive to introverts.

Therefore, in parallel to designing workplaces for different activities, we also need to create spaces that support a range of personality types. Following on from our psychoacoustic research (Oseland and Hodsman, 2017), we proposed that if the balance of personality traits in each team is known, then the appropriate acoustic environment can be created for the predominant personality type. Provide a calming base area for the more introverted teams but provide a stimulating base for predominantly extroverted teams, see figure. Of course, other activities need to be supported – introverts need spaces to occasionally interact with colleagues and extroverts to take a break from them (Figure 10.4).

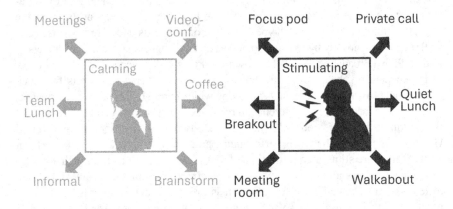

Figure 10.4 Calming or stimulating base zone (Oseland, 2022).

Interaction, collaboration and personality

I occasionally joke that you know when there are extroverts at a meeting because they tend to ask more questions and are likely to speak off the cuff, where introverts are less likely to ask questions but then later send a lengthy email raising their concerns. Following an extensive literature review of research into team performance and collaboration (Oseland, 2012), I summarised the potential implications of the *Big Five* and Myers-Briggs personality types on communication and collaboration preferences, see Tables 10.1 and 10.2.

Table 10.1 Big Five (OCEAN) implications for collaboration

Open – Prefer F2F meetings, brainstorming, plus stimulating, different and new spaces.	**Conservative** – Prefer formal, familiar, conforming and traditional spaces.
Conscientious – Prefer planned, formal, well-organised and minute meetings.	**Casual** – Prefer impromptu and informal meetings, idea generation and quick interactions.
Extroverts – Prefer F2F meetings and socialising, large social groups plus impromptu, informal, off-site meetings and stimulating spaces.	**Introverts** – Prefer written communications, distributed information, small groups, teleconferences and subdued spaces.
Agreeable – Prefer large meetings with structure and distributed information to help gain group consensus.	**Challenging** – Prefer unstructured F2F meetings where they can be more vocal and challenging.
Neurotic – Prefer well-planned formal meetings with advance notice and information; also subdued environments.	**Stable** – More comfortable with large, impromptu or informal meetings.

Table 10.2 Myers Briggs (MBTI) implications for collaboration

Extroversion – Prefer F2F and socialising, large social groups plus impromptu, informal, off-site meetings.	**Introversion** – Prefer written communications, distributed information, small groups and teleconferences.
Sensing – Prefer information and detail, plus planned and minute meetings.	**Intuition** – Prefer graphics and concepts, group brainstorms and F2F meetings.
Thinking – Prefer data and lists, plus like to challenge and discuss at meetings.	**Feeling** – People focussed so prefer, F2F informal, cosy, chatty and 1:1 meetings.
Judging – Prefer local planned, chaired and minute meetings.	**Perceiving** – Prefer local, impromptu, informal and convenient meetings.

Subsequently, the results from the literature review informed the development of an online survey, with 937 respondents (Oseland, 2013) on preferred meeting and collaboration spaces. The survey results confirm that different personality types (assessed using the *Big Five*) have different workplace preferences. For example, the survey demonstrated that introverts spend more time in solitary activity, predominantly communicate using email and when they meet, prefer enclosed offices and meeting rooms. On the other hand, extroverts spend more time in meetings, more time out of the office and less time computing than introverts. Extroverts also have higher preference for in-person face-to-face (F2F) interactions and meeting in bars, informal/social spaces and huddle rooms. Like introverts, those high in neuroticism prefer email for communicating and prefer documented information rather than group meetings for sharing information. The more neurotic types spend less time in F2F meetings and more time in solo activity, and they also dislike one-to-one (1:1) meetings for discussing personal problems.

Those higher in openness, the creative and artistic personality types, favour F2F meetings and prefer meeting in bars, huddle spaces or cafés rather than a formal meeting room. In contrast, those with a more conservative personality prefer formal meeting rooms and least like informal meeting space. Those rating higher in conscientious prefer breakout space for socialising and generating ideas, whereas those less conscientious prefer the bar/hotel or co-working space/club. Respondents scoring higher in agreeableness prefer meeting in groups for generating ideas, but they prefer intimate 1:1 meetings for socialising. Those low in agreeableness selected co-location and connectivity to the team as key design features for meeting spaces.

Desk location and personality

In my study of workplace preferences (Oseland and Catchlove, 2020), the *Big Five* was used to ascertain introversion-extroversion personality profiles. I found that extroverts had a higher preference for open plan workspaces, agile working and hot-desking, whereas introverts had a higher preference for private offices. Like introverts, those higher on the neuroticism scale also had a lower preference for hot-desking (Figure 10.5).

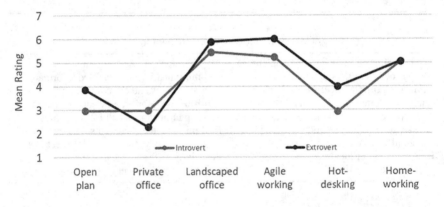

Figure 10.5 Desk location preferences by introversion-extroversion (Oseland and Catch-love, 2020).

My most recent online survey (Oseland and Raw, 2024) included 490 respondents, who had worked both at home and in their employer's offices and who self-rated their personality on an introversion-extroversion scale. The respondents also rated 51 possible influences, relating to work activities, workspace, personal benefits and sense of purpose, to indicate whether they were better supported working from home or in the office. The largest differences in the influences, between introverts and extroverts, on working in the office are shown in Figure 10.6.

Overall, extroverts are more likely to prefer working in the office whereas introverts prefer their home. This could be partly a direct effect of personality on the value placed on social interaction, and partly a consequence of the type of work undertaken by people according to their personality. This finding is in line with other research that shows extroverts are easily distracted at home and prefer the company of their work colleagues to working alone. More specifically, extroverts believe the office better supports creativity, teamwork and meetings and leadership/management. It is important to recognise that employees' age, experience and personal circumstances may also affect their preference for home or office working.

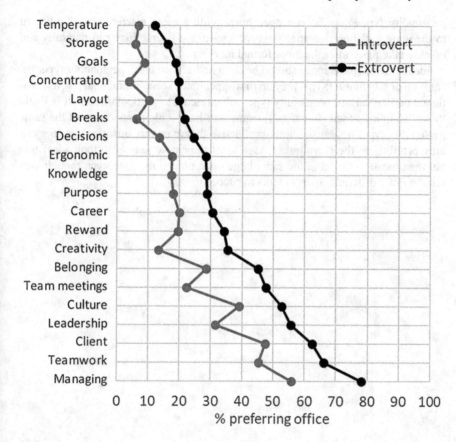

Figure 10.6 Office or home preferences by introversion-extroversion (based on Oseland and Raw, 2024).

Implications for office design

My own and other research convincingly demonstrates that personality affects the spaces where people prefer to interact and work, within and outside the office. We must provide workspaces that accommodate all personality types and suit the introverts, conscientious and more neurotic people, rather than simply build stimulating, open plan, buzzy environments that best suit extroverts and more open workers. This involves providing calmer and quieter areas for carrying out work requiring focus and concentration as well as a range of areas for different types of interaction such as 1:1 personal meetings or creative team meetings.

My landscaped office concept (Oseland, 2022) includes a variety of work settings that accommodate different personality types. For example, those more introverted, conscientious and neurotic (or sensing and thinking types on the MBTI) prefer time to think things through and develop their ideas before sharing them publicly in large formal meetings. The participation in collaboration of these personality types may be enhanced by providing more discrete spaces adjacent to the main collaboration space, where one-to-one interactions can naturally and quickly take place after the more formal meeting.

Ideally, workspace zones should be designed according to personality type, i.e. calming or stimulating. This may on first appearance seem impractical but, as mentioned earlier, certain personality types are attracted to particular roles. It is likely those discipline-based (functional) teams will have a high proportion of the same personality type. It seems counter-productive that many organisations use personality profiling at the recruitment stage to fit their new joiners to their ideal role but then locate them all in the same large open plan workspace even though they require quite different environments to succeed.

Practitioner perspective

Casey Lindberg

Introduction

This section explores the implications of employee personality differences on office design. Also considered is the design process itself and the implementation of such design strategies within the larger ecosystem of work-related spaces.

The value of targeting individual characteristics among customers and voters, such as a person's beliefs and values, within an advertising or political campaign is so widely accepted that it has become common sense. Likewise, corporations have long paid attention to the relationship between employee personality and work-related outcomes such as job performance and satisfaction, and organisations have used personality metrics to make hiring decisions for several decades to assess culture fit (Morgeson et al., 2007). However, historically, very little attention has been paid to the actual design of those organisations' workplaces regarding their employees' individual differences, including personality traits. Instead, when differences between employees are considered in workplace-related designs, the emphasis has been on differences not in their own personalities but in their job type, related work activities and adjustable ergonomic setups based on varying physical sizes and abilities. In other words, employees are often treated as interchangeable parts.

Yet, the organisational costs related to employees greatly overshadow other related costs, such as real estate investments, technologies and building operations (Kernohan et al., 1996). So, it stands to reason that if a workplace design can impact even the most modest gains in productivity, lower turnover or otherwise increase the value that an employee brings to an organisation, the investment will be worthwhile.

Traditional office layouts and personality

So, what are some examples of how personality differences may interact with different spatial solutions of office layouts? When considering the traditional types of individual enclosures seen in modern workplaces (i.e., open-office bench seating, cubicles, and private offices), there is evidence that those designs may interact with personality factors and affect self-rated outcomes. More extroverted employees have been shown to exhibit more positive outcomes like being happy in the moment, when in an exposed open-office bench seating workstation setup compared to a private office (Baranski et al., 2023). On the other hand, employees with higher levels of neuroticism, compared to those with lower levels of neuroticism, have been shown to exhibit more negative outcomes in those same more exposed workspace setups, such as having lower perceived control, and being less focused and on task (Baranski et al., 2023; Lindberg et al., 2016).

Very little research like the studies mentioned above has explored relationships between personality traits and workplace designs on meaningful, work-related outcomes. Yet even from this small amount of evidence, there are lessons learned that may have design implications related to the value of an organisation's real estate investment. In the realm of workplace design, it's important to recognise that an individual within their surrounding environment forms a dynamic system. So, a shift towards an employee-centred rather than workstation-centred process is critical.

The primary focus of workplace design should be to first support the needs of the worker. While there are monetary calculations for average workplace square footage, utilities and rent, it's important to acknowledge that there is no such thing as an average composite employee. In practice, workstation designs are implemented for distinct and unique individuals, varying in many characteristics including personality traits. And even though there are findings related to relatively stable constructs like extraversion and neuroticism, a single employee's momentary ratings on focus and happiness *throughout* a given day at work can vary dramatically. In other words, there is the need to consider not only differences between employees, but also differences *within* each employee. For example, even an employee with high levels of extraversion and low levels of neuroticism will likely vary in their ratings of feeling focused throughout the workday, regardless of their workplace setup.

An employee-driven conversation forces the designer to recognise that there is a continuum of personality-influenced outcomes. A continuum of spatial affordances in the workplace could therefore better meet the needs of a diverse workforce with varying tasks over the course of a day, compared to a uniform set of assigned workstations, cubicles or private offices. But before discussing more implications on workplace design processes, a bit of historical context is called for.

Relevant historical influences on workplace design

For decades, the open office has faced criticism based on evidence of increased distraction and lowered privacy, production and collaboration. Yet, these criticisms often neglect to pinpoint the real culprit of these negative outcomes: the open office workstation. The workstation has become all things, providing all affordances, to the open office employee. It has become an employee's default location for all types of work-related tasks, personal activities and even socialisation. In other words, its list of affordances has become untenably long (Lindberg et al., 2023). One could make a similar argument for the office cubicle. When an employee collaborates or socialises with others at their workstation or in their cubicle, they are proximate and distracting to other workers that may be trying to focus on individual work tasks or take a quiet mental break.

Recently, the pandemic has shifted the way many workers think about the role of the office. This is because advances in technology have provided flexibility in how people can access different resources including their co-workers. The result is that we now need to think about the workplace as an ecosystem of spaces (Wahi et al., 2020). When an office becomes not just a container for all things work-related, it should become a meaningful part of a system. The office, in other words, will have a distinct purpose for an organisation's goals. To be worth the investment, it should be able to support activities related to those goals better than any other part of the ecosystem.

Workplace design process implications

Consider an office that does not have a distinct purpose. It has banks of open office workstations to support all sorts of work-related activities and conference rooms to support larger meetings. Employees regularly collaborate, both in person and virtually, at their workstations. Employees with higher levels of neuroticism, and presumably higher average ratings of being able to focus in an open-office workstation due to distractions from their co-workers, may seek out a different part of that ecosystem to perform individual focused work. They may seek out home settings for those work activities to feel more focused and less distracted.

Instead, consider an office that has a more intentionally designed part of the work ecosystem. In this office location, a dense, urban environment, rent is very high and home workspaces are less available and very limited in size. One part of the office has therefore been designed to intentionally afford focused work and mental breaks, without any distractions from collaborative or social activities. In another part of the office, affordances for socialisation and brainstorming have been designed. In this office, employees with higher levels of neuroticism may thrive better on focused tasks, yet also receive the other social benefits of coming to the office rather than staying at home.

The difference between these two offices is the intentional layering of affordances that can work in concert rather than conflict with one another. Workdays are difficult to completely plan out and predict, so areas of the office need to be designed to be flexible in their intended use. But a balance must be struck between flexibility and rigidity. Too much flexibility in intended use provides less behavioural direction and leads to conflicting work activities, like open office workstations tend to afford. Too much rigidity in intended use, such as a different space for every type of activity throughout the workday, is unrealistic to effectively use and plan, and thus easy to over-utilise or under-utilise (Figure 10.7).

When affordances are intentionally layered that can work in concert, an organisation can more effectively utilise their investment as a meaningful part of the work ecosystem and provide for personality differences in the use of those spaces. Distinct parts of the office can support work activities that can be completed in the same area by teams with people with varying personality traits, like an area for quiet focused work and mental breaks, or an area that can support collaboration and socialisation. When affordances are not intentionally layered, like when an area of the office is meant to support both focused work and collaboration, distractions occur and affect those with varied levels of extraversion and neuroticism differently.

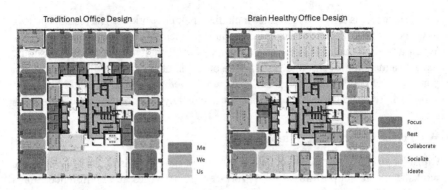

Figure 10.7 Traditional and brain-healthy office design (image courtesy of HKS).

Design communication

Of course, activity-based office designs have been around for decades. The argument here is that their affordances have not typically been layered with intention, especially to support the spectra of personalities in a workforce, and that the omnipresent open office workstation is at the heart of the problem. But even when an office space has been designed with intentional spaces for varying work activities, they often are not used according to those intentions. Why is this? The problem lies in the way the design is communicated with the organisation.

Change is difficult, and when a part of the office, such as the open office workstation, has become default for so many years for so many workers, changing work habits takes time. Adding to the problem is that an organisation's workforce is not static; even in stable companies, people retire, quit and get hired. Communication, therefore, must be ongoing, and part of the culture of the organisation.

Change enablement and change management are critical here, and often undervalued, yet largely outside the scope of this chapter. What can the designer do to increase the odds that the intentional affordances of different spaces in the office are communicated to employees? The first thing a designer can do is to document the intended usage of a space in a sort of playbook or manual for the organisation. This needs to be distributed and discussed, but this documented playbook is often never created. The playbook, which would indicate the intended, layered affordances in each space of the office and provide guidelines for which spaces should be reserved, how, and for how long, serves as a consistent starting point from which to evolve. A playbook like this may provide a more grounded direction and a sense of comfort for those with varying levels of extraversion and neuroticism.

Second, the designer can communicate the various intended uses of office spaces through materiality choices and other aesthetic decisions, thus helping create more legibility (Weisman, 1981). People's brains have a superpower of prediction using concepts depending on current context and learned past behaviour (see Yee and Thompson-Schill, 2016). This means that when a person sees a part of the office, their entire history of behaviour in similar-looking spaces influences how they predict they will use that part of the office. So, when an employee sees an open office workstation, their brain predicts all sorts of things, because they've performed so many different types of activities and experienced so many things, including being distracted, at workstations before. If a designer wants to cue a different, more intentionally layered set of affordances and behaviours, like a place to either do individually focused work or take a quiet mental break, they could utilise the way our brains predict behaviours based on learned concept associations. For instance, one could lean on our learned history of behaviours associated with libraries, theatres, quiet rail cars or even museums and provide related visual cues in that office space. In other words, when entering a part of the office that has more plush seating, more natural elements, subdued artwork or semi-enclosed areas broken up with partitions, it is easier for our behaviours to be in accordance with that space's intended affordances. Importantly, these behaviours become more and more ingrained over time the more they are used appropriately.

Taken together, the above considers the known impact of employee personality differences on office design and emphasises the need for a shift towards an employee-centred design process. Recognising the dynamic system formed by individuals and teams within their work environment, the importance of intentionally layered affordances to better accommodate the diverse needs and personality traits of a workforce is evident. Designers are urged to communicate these intentional spaces through documentation and visual cues that align with past learned associations, fostering a more effective and purposeful piece of the workplace ecosystem.

References

Allport, G. W. & Odbert, H. S. (1936). Trait-names: A psycho-lexical study. *Psychological Monographs, 47*(1), i-171.

Baranski, E., Lindberg, C., Gilligan, B., Fisher, J. M., Canada, K., Heerwagen, J., ... & Mehl, M. R. (2023). Personality, workstation type, task focus, and happiness in the workplace. *Journal of Research in Personality, 103*, 104337.

Briggs, K. & Myers, I. (1987). *Myers-Briggs Type Indicator Form G*. Palo Alto: Consulting Psychologist Press.

Broadbent, D. (1958). *Perception and Communication*. Oxford: Pergamon Press.

Cain, S. (2012). *Quiet: The Power of Introverts in a World That Can't Stop Talking*. New York: Penguin Random House.

Cattell, R. B. (1947). Confirmation and clarification of primary personality factors. *Psychometrika, 12*(3), 197–220.

Costa, P. T. & McCrae, R. R. (1992). *Revised NEO Personality Inventory (NEO-PI-R) and NEO Five-Factor Inventory (NEO-FFI) Manual*. Odessa: Psychological Assessment Resources.

Eysenck, H. J. & Eysenck, S. B. G. (1975). *Manual of the Eysenck Personality Questionnaire (Junior and Adult)*. Kent: Hodder & Stoughton.

Holland, J. (1966). *The Psychology of Vocational Choice*. Waltham, MA: Blaisdell.

Kernohan, D., Gray, J. & Daish, J. (1996). *User Participation in Building Design and Management: A Generic Approach to Building Evaluation*. New York: Architectural Press.

Lindberg, C., Fallon, E. & Davis, K. (2023). We need to break the workstation. Retrieved from: HKS Architects website: https://www.hksinc.com/our-news/articles/we-need-to-break-the-workstation/

Lindberg, C. M., Tran, D. T. & Banasiak, M. A. (2016). Individual differences in the office: Personality factors and work-space enclosure. *Journal of Architectural and Planning Research, 33*(2), 105–120.

Morgeson, F. P., Campion, M. A., Dipboye, R. L., Hollenbeck, J. R., Murphy, K. & Schmitt, N. (2007). Reconsidering the use of personality tests in personnel selection contexts. *Personnel Psychology, 60*(3), 683–729.

Nguyen, B. (2020). Personality can be the eureka moment to deciding your career. *Los Angeles Times, HS Insider*, December.

Oseland, N. A. (2012). *The Psychology of Collaboration Space*. London: Herman Miller.

Oseland, N. A. (2013). *Personality and Preferences for Interaction, WPU-OP-03*. Workplace Unlimited Occasional Paper.

Oseland, N. A. (2022). *Beyond the Workplace Zoo: Humanising the Office*. Oxon: Routledge.

Oseland, N. A. & Catchlove, M. (2020). Personal office *preferences. Proceedings of Transdisciplinary Workplace Research Conference, TWR 2020*, Frankfurt.

Oseland, N. A. & Hodsman, P. (2017). Chapter 4: Psychoacoustics: Resolving noise distractions in the workplace. In *Ergonomics Design for Healthy and Productive Workplaces*. Abingdon: Taylor & Francis.

Oseland, N. A. & Raw, G. J. (2024). *The Enticing Workplace: Attracting People Back to the Office, WPU-OP-04*. Workplace Unlimited Occasional Paper.

Schaubhut, N. A. & Thompson, R. C. (2008). *MBTI Type Tables for Occupations*. Mountain View: CPP Inc.

Wahi, N., Ramer, A., Fallon, E., Hutchison, J., Martin, M. & Lindberg, C. (2020). The future of work, Part One: Work is an ecosystem. Retrieved from: HKS Architects website: https://www.hksinc.com/our-news/articles/the-future-of-work-part-one-work-is-an-ecosystem/

Weisman, J. (1981). Evaluating architectural legibility: Way-finding in the built environment. *Environment and Behavior, 13*(2), 189–204.

Yee, E. & Thompson-Schill, S. L. (2016). Putting concepts into context. *Psychonomic Bulletin & Review, 23*, 1015–1027.

Yerkes, R. M. & Dodson, J. D. (1908). The relation of strength of stimulus to rapidity of habit-formation. *Journal of Comparative Neurology and Psychology, 18*, 459–482.

11 Supporting the neurodiverse

Editors' introduction

People are increasingly knowledgeable about their own neurodiversity and working to attain design-related support at their workplaces so that they can work to their full potential. Empathetic colleagues back these efforts as do organisations seeking to elevate the performance of all employees.

Supporting a single neurodivergent person in a home is significantly different than supporting people with assorted neurotypes working together at a single location. Space planning is made easier by the fact that people in the workforce are unlikely to be at the extreme of neurodiversity continuums. Also, neurodivergent workers are likely to have had a range of previous life experiences that have taught them how to cope with their conditions. For example, how to best restore their psychological state if an environmental stressor, or something similar, has unduly affected them.

Well-designed workspaces should support the wide range of neurotypes in the workforce today.

DOI: 10.1201/9781003390848-11

Researcher perspective

Kristi Gaines

Neurotypes

There is an increased desire to become a more inclusive society. To do so means addressing the needs of the many, not just the few. And that includes addressing the needs of the growing number of neurodivergent individuals in the workplace. While the functioning of neurotypical individuals falls within set norms, neurodivergents (or neurominorities) fall outside of those parameters. Neurodivergence addresses the unique way we all think, feel and act that is considered different to the predominant neurotype. Those who are neurodivergent are born with a neurological difference that typically remains for their entire life. Though neurodivergence can occur after a trauma or injury, those are a small percentage of cases. According to Understood (2022), 70 million people in the United States, including one in five children, have learning and thinking differences, such as dyslexia and attention-deficit hyperactivity disorder (ADHD).

As our awareness related to neurodiversity grows, so does the interest in becoming a more inclusive society. Neurominorities differences can be an extraordinary strength in the workplace. Neurodivergent thinkers often possess exceptional talents when it comes to innovation, creative storytelling, empathy, design thinking, pattern recognition, coding and problem-solving. Ensuring we approach the design of workplaces to be welcoming for all, including neurominorities, is key to addressing social inclusion and to more successful organisations. In recent years, a growing number of companies have begun to realise that accommodating the different wiring of neurodiverse people can provide a huge competitive advantage that is good for business. Especially in times like we are facing now – with an increased "war for talent", labour shortages and an increased need for innovative thinking – welcoming the neurodiverse population into the workplace has never been more critical. This is leading to a range of more inclusive policies, programs and procedures, though this recognition is only just beginning to affect workplace design.

Neurodistinctions

While differences in the way we process information and sensory stimulation can present challenges, disablement results when there is a mismatch between individuals' needs and their environments (Olkin, 2022). Some neurodivergent people, particularly those with autistic spectrum disorders (i.e. on the autism spectrum), may be particularly sensitive to the surrounding environment primarily because of sensory processing, integration and modulation difficulties (Gaines et al., 2016).

Depending on their response to sensory stimuli, some neurominorities may be considered hypo- or hypersensitive (Figure 11.1).

- *Hyposensitive* individuals often have difficulty seeing, hearing or feeling the acute sensory details in a given environment. They tend to prefer to be over-stimulated and need more stimuli to successfully process sensory information.
- *Hypersensitive* individuals process the details of sensory stimuli in an overly magnified way. They prefer predictable environments with controlled stimuli. They tend to dislike environments with excessive stimuli such as bright lights, crowds, unfamiliar scents, textures or temperature fluctuations.

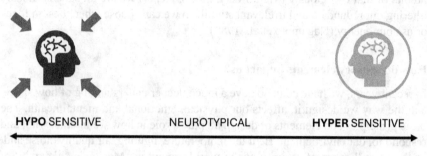

HYPO SENSITIVE   NEUROTYPICAL   **HYPER** SENSITIVE

- Difficulty perceiving acute sensory details
- Prefer to be over - stimulated
- Need more stimuli to process information

- Amplify details of sensory stimuli
- Prefer control over stimuli
- Dislike excessive stimuli

Figure 11.1 The hypo- to hypersensitivity range.

Research suggests that sensory processing sensitivity can accompany neurodivergence. For example, Samson (2021) explains that "both autism and high sensitivity are examples of neurodivergence"; however, highly sensitive processing is not a form of autism per se and "autistic individuals may have either a hyper- or hypo-reactivity to sensory information, a combination of both, or neither." Considering other neurominorities, Scott (2024) clarifies that while highly sensitive people (HSPs) and those with ADHD both "exhibit over-responsiveness to stimuli, people with ADHD also exhibit cognitive symptoms that HSPs do not such as difficulty focusing or paying attention."

Different neurological conditions, or neurodistinctions, manifest in different ways, and even people who share the same condition may experience it to varying degrees and express it in different forms. "If you've met one person with autism, you've met one person with autism," said Dr Stephen Shore, an advocate for neurodivergents (cited by Margo, 2024). We need to acknowledge that we all have different thresholds and tolerances for handling sensory input, and that includes neurotypicals who are not immune to sensory stimulation Hence, we need to create environments that support the various needs of individuals throughout their day so we can all be effective and thrive, which is good for us personally and for business.

Though neurodiversity is often an invisible disability, neurominorities make up one of the largest underrepresented groups in the workplace (Haynes, 2022), so it's imperative that we create neuro-inclusive work environments. Neurodivergents often possess a different skill set that can be invaluable. For example, individuals with ADHD demonstrate higher levels of divergent thinking and the ability to "thinking outside the box." We need team members who have acute perception, insight and honesty. They are essential to discover, innovation and high-functioning teams.

Humans are sensory beings, and we are constantly receiving sensory input. It has been documented that on average, we receive 11 million bits of sensory information every second but are only capable of processing a minute fraction, about 50 bits of that consciously (Markowsky, 2024). Hence, we are all subconsciously filtering out redundant and irrelevant stimuli so we can choose where best to focus or use our energy (Nagamoto et al., 1989).

How do sensory elements impact us?

In recent years, we have come to have a much deeper understanding of how space, and the way we design it, affects our physical, emotional and mental health. The basic principles and elements of design all play a role in how we interact with, and respond to, our environment. Be it the temperature, lighting, air quality, noise and/ or the overall sense of security of the world around us. Maslow's "hierarchy of needs" indicates that if the physiological and safety needs of an individual are not met, it is difficult for them to perform other tasks and their engagement and wellbeing will suffer (Figure 11.2).

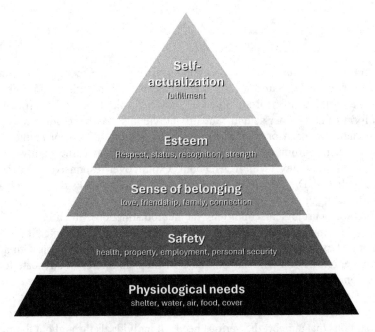

Figure 11.2 Maslow's hierarchy of needs.

Our sensory system is made up of multiple senses including the following generally accepted ones in neuroscience research.

* *Auditory* – Worldwide studies consistently show that noise levels are a major cause of dissatisfaction in the workspace (e.g. Leesman, 2022). Acoustics includes general sound levels, sound transmission, echoes and reverberation. Many neurominorities suffer from misophonia, a severe sensitivity to specific sounds, such as chewing or crunching paper. Excessive noise, chatter of others on calls and other unwanted sounds can be a major distraction. While many are bothered by sound, or lack of it in shared spaces, neurominorities can find soundscapes downright disabling. The acoustic level across a space should vary to reflect the activity level and function needs of the space. For example, social spaces may have a higher level of lighting, energy and sound versus spaces that are designs for focus and concentration. Unexpected sounds have been identified as the most problematic sensory trigger for people who have sensory processing differences.

  Effective acoustic design for the workplace provides a variety of auditory settings in support of diverse activities, locates them appropriately relative to one another and specifies attributes for acoustic comfort within spaces and acoustic separation between them. Acoustic design may also consider whether a sound masking or white noise system would further increase comfort. And many neurodivergents and neurotypicals alike take personal control and use noise-cancelling headphones. Thoughtful material selection may be one of the primary ways that an interior designer can control noise in existing buildings where layout and construction will not be altered. Designers may address unexpected sounds by insulating loudspeakers and defining areas according to activity and purpose (Gaines et al., 2016).

* *Visual* – Our visual sense is the primary way we perceive the spaces we occupy and plays a major role in our we interact with our environment. In fact, 80% of our sensory input comes through our sight (Rosenblum, 2010). For some, having additional visual stimulation is key to ensuring they are invigorated while others find high levels of visual information very distracting.

  Creating options for in increased visual stimulation in social or transitional spaces can aid in wayfinding and providing relief and a boast of energy from the more calm and controlled spaces where people are doing more focused or concentrative work. Selecting colours, patterns and materiality are all important elements in providing visual control and relief. For some, additional screening to limit distractions may be needed, while others thrive on the visual energy of being in shared spaces – also considering providing signage and information in a variety of ways to communicate with various processing preferences.

* *Olfactory* – Our sense of smell, or olfaction, is the sense through which smells are perceived and links to the memory and emotional functions of the brain. Smell impacts our ability to detect desirable foods, hazards, and pheromones and plays a role in taste. Many individuals have hypersomnia, an increased sensitivity to smell. Our sense of smell runs straight to our limbic system, the

region in our brain that regulates emotions and memory. A specific smell can help conjure up a childhood memory or remind us of a specific brand and how it wants to be perceived. People working in offices with enhanced ventilation have been found to score higher in their cognitive function when responding to a crisis or developing a strategy than those working in conventional environments (Allen et al., 2016).

Scenting can be extremely divisive and must be approached in a subtle manner, if at all. As a rule, try to ensure there is good ventilation to reduce negative smells and ensure you use low-volatile organic compounds (VOCs) in materials and furniture. Consideration should be given to location areas that generate strong smells, such as pantries in enclosed spaces. The strategic use of aroma in support areas that individuals can elect, or not, to use, such as refresh rooms, etc. will provide enhanced environmental conditions for the users

- *Gustation* – Our sense of taste is linked to our sense of smell. Gustation can affect our ability to focus. Studies have found a positive correlation between mindfulness and healthier eating. Healthier eating results in better overall health, which can lead to lower absenteeism.

  Encouraging people not to eat at their desks and creating more mindful spaces for eating could improve health in terms of encouraging healthier attitudes towards food for everyone.

- *Tactile* – Our sense of touch is one of the main ways we interact with our surroundings. Specifically, individuals with ADHD seek out more tactile stimulation but less auditory and visual input. People often simultaneously know how something will feel based on visual information. Many hyposensitive individuals need physical interaction with their environments and need to engage with their surroundings.

  Related to tactile is personal space which refers to the distance that people place between themselves and others and are influenced by social values. The theory of proxemics developed by Edward T. Hall (1963) explains four basic zones in interpersonal distance of intimate (<1.5', <0.5m), personal (1.5'–4', 0.5–1.2m), social (4'–12', 1.2–3.7m) and public (>12', >3.7m). Neurodivergents frequently display differences in personal space than the general population. For example, personal space may be larger for individuals on the autism spectrum who will need a larger space in the company of unfamiliar people and places (Gaines et al., 2016).

  Consider creating low-to-no touch spaces for those who are hypersensitive to touch while providing areas that encourage physical touch and interaction for those who are hyposensitive, or desire increased tactile engagement. Break-out spaces, wider transition spaces and solitary seating areas will allow for comfort and productivity.

- *Proprioception* – Our sense of positioning and movement of the body and interceptive is our internal sensing or signals from our bodies.

- *Vestibular* – Refers to our equilibrium and balance can be impacted by elements within a space. The built environment and the elements within it impact centring and sense of balance with our surroundings. Individuals may have difficulty with changes in surfaces and may become disoriented.

To aid proprioception and the vestibular sense, designing circulation paths and general open spaces with clear lines of sight, ample spacing and limited distractions can reduce the struggles of individuals that have challenges with proximity, density and balance. Minimise floor material and elevation changes, allow for banisters and supportive devices and allow space for movement.

- *Interoception* – Our sense of internal body awareness where our internal systems, e.g. digestion and breathing, inform us of our bodies' performance and any stress we may be experiencing.

Thermal comfort falls between tactile sensation and interoception. Along with acoustics, and visual stimulation, thermal comfort consistently ranks on workplace surveys as one of the top environmental irritants. Numerous researchers have found that thermal comfort can increase productivity (see review by Oseland, 2022). Thermal comfort can vary with personal factors such as clothing, activity level and metabolism, as well as neurology.

Where possible provide individual temperature controls, and elements such as an operable window or air diffuser, to enable workers to adjust their thermal environment to their liking. An alternative option is to leverage environmental sensors that can identify various temperature levels within the space and enable individuals to relocate as desired.

Practitioner perspective

Nigel Oseland and Sally Augustin

Why design for neurodiversity?

The Chartered Institute of Personnel and Development (CIPD), the UK's human resources professional body, estimates that neurominorities represent around 10% of people in the workforce such "that more than one in ten job applicants, existing staff and customers are likely neurodivergent in some way" (CIPD, 2018). Many organisations, particularly in the technology and engineering industries, have recognised the highly developed cognitive skills that neurodivergent workers bring to their organisation. Indeed, "When Microsoft and others hire neurodivergent people in front-line roles, it's a clear statement that such organisations have already recognised that building neurodiverse teams can be a competitive advantage and is part of being a responsible employer." (CIPD, 2018). A tour, made by Oseland, of Microsoft's offices in Seattle several years ago highlighted the specialist facilities provided, such as restorative rooms, and the processes in place to accommodate such neurodiverse employees. In contrast, the number of UK employment tribunal cases related to neurodiversity is increasing, with neurodiverse individuals claiming they are being unfairly treated as allowances are not made for their condition, including workplace layout issues (Cox, 2024).

Oseland (2022) argues that the workplace industry "needs to design for the range not the average" and moves away from homogenous layouts and a one-size-fits-all mentality. The requirements for accommodating neurominority workers may not be as challenging as initially anticipated. Those in gainful employment are unlikely to be at the extremes of the neurodiversity spectrum as they usually attended school and learned how to cope and restore their own mental equilibrium after disruption. Indeed, Clouse, Wood-Nartker and Rice (2020) report that "recommendations demonstrate that sensitivity to the needs of people with autism creates a solution that is better for all people." Kay Sargent and colleagues at HOK expand on this principle and propose:

> designing for the neurodivergent creates space that enables all individuals to find suitable levels of privacy and concentration, connection and engagement … The significant overlap between design for neurodiversity and the wider movement toward design for improved health and wellbeing suggests that the benefits of a more inclusive workplace apply to the entire population.
>
> (HOK, 2019)

Design recommendations for accommodating the neurodiverse

Design scope

The British Council for Offices (BCO) reports that "Design is only one element of an enabling work environment; we must also consider the culture ... The physical office space is part of a wider employment ecosystem" and "A key principle of making an environment enabling is to provide people with autonomy through choice and a variety of physical amenities" (BCO, 2022). The CIPD (2018), for example, focuses on organisational, culture and leadership initiatives to better accommodate neurodiverse workers.

The focus of this chapter, however, is on workplace design, particularly, the environmental conditions, interior design and internal layout as per the previous chapters, see Figure 11.3. The British Standards Institution (BSI, 2022), BCO (2022) and Maslin (2022) all provide design guidance, to accommodate the neurodiverse, for the building site and structure, including its façade, windows, entrances/exits, stairwells, reception, lobbies and surrounding landscape. BSI also stresses the importance of consulting neurodiverse occupants on their requirements, but some caution is required in selecting participants for engagement and respecting their privacy.

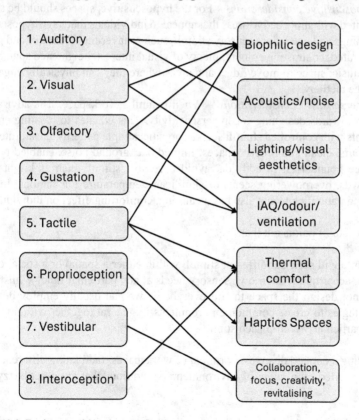

Figure 11.3 Design considerations to facilitate sensory responses.

Sensory differences

A key aim of BSI's (2022) guide to designing for neurodiversity is to influence the design and management of the built environment to reduce the potential for sensory or cognitive overload or distress from features within it. The design recommendations that follow relate to hypo- and hypersensitivity, discussed earlier, along with guidance for specific neurodiversities. Steve Maslin, an architect and long-term evangelist for inclusive design, and Kay Sargent, director of thought leadership at HOK, both highlight that the neurodiverse can find certain designs and environmental factors, such as lighting, sound, odours, temperature, privacy and security, over-stimulating, distracting, confusing and stressful or in some cases be under-stimulating (HOK, 2019; Maslin, 2022).

HOK's (2022) survey of 202 neurodiverse individuals found that almost one-quarter (22%) of respondents considered themselves hyposensitive overall, whereas almost one-half (47%) considered themselves hypersensitive. However, some individuals were found to be hyposensitive to some sensory stimuli while being hypersensitive to others. Sargent and colleagues at HOK[1] propose the following overarching design guidance.

- *Hyposensitive considerations* – For the hyposensitive, spaces should be supplemented with more active areas that appeal to individuals that have hyposensitive leanings and need more sensory stimuli, such as layering of textures and planes; saturated, contrasting colours; plenty of visual interest; heightened sound levels or music; space to move/fidget and areas where they can physically engage the space tactilely.
- *Hypersensitive considerations* – When planning workspace, it is advisable to lean towards designing for hypersensitivity. This equates to creating environments with controlled stimuli such as organic, simple patterns, unsaturated light colours, clean and orderly spaces, limited background noise, enabled personal space boundaries. In addition, avoid excessive stimuli such as bright lights, crowds, overpowering scents or significant temperature fluctuations. Incorporating natural elements also has a calming, comforting effect on individuals.

Maslin (2022) points out that:

People will seek out different stimuli within either a spatial or a social context, according to their own sensory needs at any particular time. Whilst we cannot design the space to suit everybody, we can usually employ design strategies to create common environments in general together with a range of particularly well-placed stimuli.

Providing a curated diverse assortment of workspaces, facilitating different activities supported by a range of environmental conditions that are quiet/buzzy, have

---

1 Personal correspondence with and advice from Kay Sargent at HOK.

bright/dim lighting, calm/ stimulating colours and include social/private spaces, for example, can particularly benefit those on the autism spectrum and those with ADHD (Ahrentzen & Steele, 2009; BCO, 2022). BSI (2022) proposes providing flexibility, choice and control to meet a spectrum of requirements and offering places for recovery and respite when needed.

Breaking up large areas into zones for specific sorts of activity is therefore desirable (Mostafa, 2020). Activity-based working environments provide such as range of spaces. However, agile working environments with unassigned desking (in essence hot-desking) will not be popular for those neurodiverse people who require a dedicated desk that they can set up, perhaps customise, in a location that best suits their preferences.

Auditory sense, acoustics and noise

Sensory distractions such as noise are undesirable in generable, but especially for the neurodiverse (Shabha & Gaines, 2011; Willis & Cross, 2022). Noise that "spills" from one area into an adjacent one is stressful for those on the autism spectrum (Willis & Cross, 2022). Noise from other areas may lead to curiosity and distraction for those with ADHD.

It is important to holistically access the stresses present in any workplace, as some stimuli may be particularly distracting to people with ADHD but not to others. For example, a squeaky chair that someone without ADHD may be able to ignore may capture and hold the attention of someone with ADHD and impede their professional work until it is eliminated. Providing an assortment of work areas can be particularly handy for people with ADHD as it is for people on the autism spectrum because options for workspaces allow them to relocate away from distractions. Maintaining heating and ventilation systems in comfortable, productive ranges is not only good for all employees but eliminates a potentially riveting distraction with the potential to capture and hold the attention of workers with ADHD. Adults with ADHD may be particularly susceptible to the negative consequences of audio distractions (Pelletier et al., 2016).

Halbesleben, Wheeler and Shanine (2013) explain that people with ADHD are likely to be distracted and recommend developing "quiet and organized work areas to help employees with ADHD focus on important tasks." BCO (2022) proposes "having 'library'-style rooms as third spaces across the building that can be accessed by all but have an agreed work culture" such as heads-down work with phones on silent, zero phone calls and minimal talking.

If such quiet areas are not practical, then at minimum ensure rooms for solo, focused work and restoration are provided, ideally in the same location on each floor. "Escape areas" that people on the autism spectrum can access when overstimulated make large offices seem more hospitable (Mostafa, 2008; Willis & Cross, 2022). These spaces need to be well-shielded acoustically and visually. The BCO recommends that such spaces are private, have a door, are accessible from the main circulation spaces without having to go through office space, are calming with modest design and muted colours, have textures and materials that are relaxing and natural, including biophilic elements that have restorative qualities. Furthermore, the BCO and BSI suggest that quiet/restorative spaces are adaptable to personal requirements with control over lighting, temperature, visual privacy and furniture layout. Sargent and colleagues at HOK (2019) also propose that this sort of room should allow neurodiverse individuals to listen to their preferred sounds or to take whatever steps are required to return to a conducive mindset.

Vitoratou et al. (2023) estimate that 18% of the UK general population may have misophonia, an extreme emotional reaction to certain everyday sounds, particularly eating, throat clearing, keyboard tapping and rustling paper. Therefore, provide spaces where people can eat away from their desks and where they can avoid hearing others eating. The BSI (2022) note that:

> Some people require a quiet space in which to be still, whilst others who are hyposensitive prefer a degree of activity or stimulation ... quiet space should have the flexibility to also be used for providing some active multisensory stimulation for people who require this where a separate multisensory environment cannot be provided.

Based on their survey and research, the HOK team also found that some occupants need a higher level of auditory stimulation, so they suggest creating some spaces that have a higher energy level or buzz, not only for social gatherings but even for some individuals to work in.

Visual sense, lighting, colour and visual aesthetics

Regarding visual clutter, moderate to low visual complexity is particularly important for people on the autism spectrum (Libassi, 2009; Remington et al., 2009) and in workplaces to be used by people with ADHD (Halbesleben, Wheeler & Shanine, 2013). Cooler coloured lights seem to enhance the cognitive performance of people with ADHD (Amor et al., 2014). Maslin (2022) warns that certain lighting patterns and other stimuli can trigger migraine and epilepsy.

The BCO (2022) suggests that if a building/floor becomes busier in activity, then its design should become more muted through aspects such as dimmer lighting and calming colour palettes. Furthermore, "question whether a space genuinely needs large overhead lighting" and "if there are high levels of natural light, use design applications that diffuse and/or control at source."

Research also shows that shadows generate stress for people on the autism spectrum, so any sort of lighting scheme that generates them is problematic (Becchio, Mari & Castiello, 2010; Willis & Cross, 2022). However, adding fill-in type lighting to eliminate potentially challenging shadows can increase the comfort of people on the autism spectrum. Natural lighting is desirable, but it is particularly important that it does not generate glare (Mostafa, 2020; Shabha & Gaines, 2011).

HOK (2019) report that some people, especially those with sensory processing challenges, have a heightened sensitivity to visual stimulation and find it distracting and sometimes confusing. For example, they can be distracted by large areas of bold geometric patterns, such as stripes, bars, solid blocks and perforated materials, so avoid such patterns in spaces where people are working for extended periods. Maslin (2022) notes that some occupants suffer from trypophobia, the fear of holes, so avoid such patterns. Furthermore, HOK explain that floor surfaces that are matte or low sheen are preferred as shiny surfaces can appear to be slippery causing confusion for those with poor sight or sensory processing challenges.

HOK (2019) also propose creating non-stimulating colour schemes intermixed with areas of higher stimulation. Muted and less saturated colours and those that occur in nature tend to be more calming and cause less sensory overload than vivid and saturated darker tones.

Olfactory sense, gustation, indoor air quality, ventilation and odours

Good ventilation is key in areas that produce odours, such as kitchens and toilets, to prevent annoyance and distraction to the olfactory hypersensitive and general building population.

Some organisations have experimented with infusing different scents into the workplace at varying times of the day to improve performance. However, Oseland (2022) explains that this is not a good strategy as some people, like those with olfactory hypersensitivity, have a negative reaction to odours that others may find pleasant such as fragrances and scents.

Tactile sense, interoception, haptics and thermal comfort

Like colour and patterns, texture can dial up or down the intensity of stimulation in an environment. HOK (2022) found that some individuals with ADHD may seek out more tactile stimulation but less auditory and visual input. They recommend creating some richer tactile experiences or texture within spaces that individuals can engage with, for example heavier pile or sculptured rugs that are strategically placed throughout office spaces.

Maslin (2022) suggests that designers consider furniture that "moves," as some people, such as those with ADHD, like to rock, lean back and stand as they work. HOK (2019) also suggest facilitating doodling/drawing in collaboration areas with white walls etc.

As mentioned in the previous section, provide individual control of thermal environments where practical. Oseland (2022) explains that in open plan work-spaces, this may be achieved using environmentally responsive workstations, which have local controls for ventilation and temperature, or by offering a voting system for the zonal set-point temperature, or by offering different work areas with a range of temperatures that occupants choose to sit in. The latter, different working environments at different temperatures to allow people to choose the most comfort-able area for them, is also proposed by the BSI (2022).

The BCO (2022) reports that a cold surface can be biologically uncomfortable for someone with a hypersensitivity to temperature changes. BCO (2022) suggests avoiding using thermally conductive materials on elements that will need to be touched, such as bannisters. The BCO also encourages designers to, where possible, use natural materials such as wood that are gentle to the touch and have biophilic qualities.

Vestibular sense, proprioception, space and wayfinding

Spatial transition zones, including windows on doors separating one space from another, reduce stress levels as people on the autism spectrum move through a space (Ahrentzen & Steele, 2009; McAllister & Maguire, 2012; Willis & Cross, 2022).

To ensure the comfort of the neurodiverse, the BCO (2022) cautions against making corridors too long and narrow – ensure that there is space to pass someone else comfortably. Likewise, the BSI (2022) suggests avoiding long narrow corridors and breaking hallways up using windows, intersections and recesses. Maslin (2022) recommends that designers be aware of vertigo, agoraphobia and claustrophobia, at bridges and ledges near large open window spaces or tightly enclosed spaces.

Spaces used by people on the autism spectrum need to be uncomplicated to navigate (Pellicano et al., 2010). The BCO recommends there should be no physical obstructions, such as clutter, steep stairs, a lack of lifts or slippery surfaces that impedes travel through a workplace. Navigation also relates to the amount of space and density, especially for those with dyspraxia or Meniere's disease, so it is important to ensure spaces are sufficient for people to circulate without encroaching upon personal space boundaries (BSI, 2022). Likewise, there should be enough space for a person using physical aids to move easily, or to move their body freely and safely – narrow paths or hallways, for example, can obstruct movement. The BCO and BSI both recommend providing clear alternative routes that allow the option to bypass particularly busy spaces, corridors and lobbies, thus enabling people to have some control over avoiding the most crowded routes.

People on the autism spectrum are more likely to prefer rectilinear spaces and design elements (Palumbo et al., 2022; Vartanian et al., 2021). Also, individuals on the autism spectrum prefer familiar to novel design elements (Vartanian et al., 2021). Carefully curated sets of space use options should be presented to people on the autism spectrum (Ahrentzen & Steele, 2009).

The BSI (2022) recommends that "people should not be forced to sit in the middle of a large space with their backs to an activity or to people moving around which triggers anxiety" and distraction for those hypersensitive to movement or with ADHD. The location of desks for the neurodiverse should be carefully considered.

Maslin (2022) states that avoiding confusion in the workplace is a key design objective. He suggests, providing key locations on circulation routes with a sense of place and making doorways look like an entrance, otherwise they may not be perceived as a portal to an environment.

Sargent and the HOK team found that for those with proprioceptive and vestibular challenges, their sense of positioning, equilibrium and balance can be impacted by design elements within a space. For example, stripes or strong colour contrast on the floor can be read as stairs or a change in levels and solid blocks can be misread as voids or holes, especially for some neurodiverse people. Likewise, the BSI (2022) suggests that "Clear sight lines and/or ease of identifying the locations of entrances and exits should be provided to reduce the potential for sensory overload and to ease orientation and movement to, from and within the building." Furthermore, designs should avoid different tonal and textural ground surface or floor finishes as they are a challenge for some resulting in tripping but different types of floor surface to distinguish circulation routes, or the colour coding of floors or amenities, can enhance wayfinding.

The BCO (2022) proposes keeping signage simple, clear and muted to avoid unnecessary confusion and stress. Offer wayfinding markers adjacent to doors/entrances, not just on them, to set a sense of place/context. The BSI (2022) recommends that wayfinding tools are provided to support at least two senses, including visual, audible or tactile methods. The BCO also recommends that designers use personal digital wayfinding aids and digital walkthroughs, rather than printed or on-wall maps which can be very confusing. Furthermore, the BCO suggests that a neurodivergent employee should be given the option to have a digital walkthrough of a new work environment, so that they can plan their journey and create their own landmarks for easier navigation. Digital twins or computer-generated imagery (CGI) renders can be used to produce a digital walkthrough of the building – the main office floors and common areas. Such a provision would benefit all employees.

Biophilic design

The benefits of biophilic design and access to nature for all occupants are explained in Chapter 2. However, it is particularly important for people with ADHD that there be access to nature at work. For example, views of green spaces have been tied to enhanced attention among those with ADHD (Kuo & Taylor, 2004).

The BCO (2022) guidance for accommodating the neurodiverse suggests providing designated quiet outdoor spaces away from busy areas that have shelter against the weather. Even smaller, older buildings may have a courtyard or small outdoor space or consider creating a mini green oasis that is designed as a calm space. BSI (2022) notes that such spaces provide the opportunity to escape from overwhelming spaces or crowded buildings to a place where personal spaces are provided.

The BSI suggests that when creating a garden for wellbeing and restorative purposes: (i) choose a location away from extraneous noise, (ii) ensure branches that scratch or creak are cut back regularly, (iii) provide large areas of shade, (iv) provide spaces to rest and reorientate when transitioning from one type of area to another, (v) create open sightlines to reduce confusion, (vi) provide places that are stimulating and spaces that are calming and (vii) avoid plants that are toxic, have sharp foliage or thorns or are strongly scented.

Conclusion

When environments are not designed to be supportive of sensory processing, individuals are forced to find ways to cope. For neurominorities, this is too common an occurrence and they often must deal with environments that are either overwhelming to them or in some cases do not provide sufficient stimulations – both can be physically and mentally taxing and degrade their professional performance. Providing workspaces that offer choice and some degree of control is more likely to result in a high-performing and empowered workplace, for all occupants and especially neurodiverse workers. Offering control over of the level and type of sensory experiences that occupants are exposed to helps create high-performance spaces that are human-centric and neuro-inclusive.

The material in this chapter makes it very clear that workplace design that supports the neurodiverse adheres to the principles of neuroscience-informed workplace design discussed throughout this book. Workplaces where the neurodiverse thrives and performs to their full potential are places where neurotypicals do as well.

Acknowledgement

The authors are extremely grateful to Kay Sargent, senior principal and director of thought leadership of HOK, for her generous advice and substantial input to this chapter on designing offices to accommodate neurodiversity.

References

Ahrentzen, S. & Steele, K. (2009) *Advancing Full Spectrum Housing: Designing for Adults with Autism Spectrum Disorders*. Phoenix: Arizona State University.

Allen, J. G. et al. (2016) Associations of cognitive function scores with carbon dioxide, ventilation, and volatile organic compound exposures in office workers: A controlled exposure study of green and conventional office environments. *Environmental Health Perspectives, 124*(6), 805–812.

Amor, C., O'Boyle, M., Pati, D. & Pham, D. (2014) Use of neuroscience in design: A comparative analysis of neural and behavior outcomes. In Carney, J. and Cheramie, K. (Eds.), *Building with Change, Proceedings of the 45th Annual Conference of the Environmental Design Research Association*. New Orleans: EDRA, 391.

BCO (2022) *Designing for Neurodiversity*. London: British Council for Offices.

Becchio, C., Mari, M. & Castiello, U. (2010) Perception of shadows in children with autism spectrum disorder. *PLoS ONE, 5*(5), e10582.

BSI (2022) *PAS 6463:2022 Design for the Mind – Neurodiversity and the Built Environment – Guide*. London: British Standards Institution

CIPD (2018) *Neurodiversity at Work*. London: Chartered Institute of Personnel and Development.

Clouse, J., Wood-Nartker, J. & Rice, F. (2020) Designing beyond the Americans with Disabilities Act (ADA): Creating an autism-friendly vocational center. *Health Environments Research and Design Journal, 13*(3), 215–229.

Cox, E. (2024) Key employer lessons from 2023 neurodiversity case uptick. *Law360*. Retrieved from: https://www.law360.com/articles/1780313/key-employer-lessons-from-2023-neurodiversity-case-uptick

Gaines, K., Bourne, A., Pearson, M. & Kleibrink, M. (2016). *Designing for Autism Spectrum Disorders*. New York: Routledge.

Halbesleben, J., Wheeler, A. & Shanine, K. (2013) The moderating role of attention-deficit/hyperactivity disorder in the work engagement-performance process. *Journal of Occupational Health Psychology, 18*(2), 132–143.

Hall, E. T. (1963). A system for the notation of proxemic behavior". *American Anthropologist. 65*(5): 1003–1026.

Haynes, A. (2022). Neurodiversity: The little-known superpower. Korn Ferry Institute. *HOK, News + Events*. Retrieved from: https://www.hok.com/news/2022-04/survey-neurodiverse-employees-workplaces/

HOK (2019). Designing a neurodiverse workplace. *HOK, New York*. Retrieved from: https://www.hok.com/ideas/publications/hok-designing-a-neurodiverse-workplace/

HOK (2022) HOK, Tarkett survey offers new insights for neurodiverse workplaces. *HOK News + Events*. Retrieved from: https://www.hok.com/news/2022-04/survey-neurodiverse-employees-workplaces/

Kuo, F. & Faber Taylor, A. (2004) A potential natural treatment for attention-deficit/hyperactivity disorder: Evidence from a national study. *American Journal of Public Health, 94*(9), 1580–1586.

Leesman (2022) *The Leesman Review, Issue 32*. London: Leesman.

Libassi, P. (2009) Formulating a plan: Developing the Debra Ann November Wing of the Lerner School for Autism at Cleveland Children's Hospital Center for Autism. *Healthcare Design, 9*(7), 46–55.

Margo, K. (2024) Autism is one word attempting to describe millions of different stories. *Autism Speaks*. Retrieved from: https://www.autismspeaks.org/blog/autism-one-word-attempting-describe-millions-different-stories

Markowsky, G. (2024) Information theory. *Encyclopaedia Britannica*. Retrieved from: www.britannica.com/science/information-theory/Physiology

Maslin, S. (2022) *Designing Mind-Friendly Environments: Architecture and Design for Everyone*. London: Jessica Kingsley Publishers.

McAllister, K. & Maguire, B. (2012) Design considerations for the autism spectrum disorder –Friendly key stage 1 classroom. *British Journal of Learning Support, 27*(3), 103–112.

Mostafa, M. (2008) An architecture for autism: Concepts of design intervention for the autistic user. *International Journal of Architectural Research, 2*(1), 189–211.

Mostafa, M. (2020) Architecture for autism: Built environment performance in accordance to the autism ASPECTSSTM design index. *Autism, 360*, 479–500.

Nagamoto, H. T., Adler, L. E., Waldo, M. & Freedman, R. (1989) Sensory gating in schizophrenics and normal controls: Effects of changing stimulation interval. *Biological Psychiatry, 25*(5), 549–561.

Olkin, R. (2022) Conceptualizing disability: Three models of disability. *American Psychological Association*, March. Retrieved from: https://www.apa.org/ed/precollege/psychology-teacher-network/introductory-psychology/disability-models

Oseland, N. A. (2022) *Beyond the Workplace Zoo: Humanising the Office*. Oxon: Routledge.

Palumbo, L. et al. (2022) Visual preferences for abstract curvature and for interior spaces: Beyond undergraduate student samples. *Psychology of Aesthetics, Creativity and the Arts, 16*(4), 577–593.

Pellicano, E. et al. (2010) Children with autism are neither systematic nor optimal foragers. *Proceedings of the National Academy of Sciences, 108*(1), 421–426.

Pelletier, M.-F., Hodgetts, H., Lafleur, M., Vincent, A. & Tremblay, S. (2016) Vulnerability to the irrelevant sound effect in adult ADHD. *Journal of Attentional Disorders, 20*(4), 306–316.

Remington, A., Swettenham, J., Campbell, R. & Coleman, M. (2009) Selective attention and perceptual load in autism spectrum disorder. *Psychological Science, 20*(11), 1388–1393.

Rosenblum, L. D. (2010) *See What I'm Saying: The Extraordinary Powers of Our Five Senses*. New York: W.W. Norton & Company.

Samson, R. (2021) No, being autistic is not the same as being highly sensitive. *Psychology Today.* Retrieved from: https://www.psychologytoday.com/gb/blog/the-highly-sensitive-child/202112/no-being-autistic-is-not-the-same-being-highly-sensitive

Scott, E. (2024) Are you a highly sensitive person? When sensitivity rises to the level of neurodivergence. *Verywellmind.* Retrieved from: https://www.verywellmind.com/highly-sensitive-persons-traits-that-create-more-stress-4126393

Shabha, G. & Gaines, K. (2011) Therapeutically enhanced school design for students with autism spectrum disorders (ASD): A comparative study of the United States and the United Kingdom. In Mittleman, D. & Middleton, D. (Eds.), *Make No Little Plans, Proceedings of the Annual Meeting of the Environmental Design Research Association.* Washington: Environmental Design Research Association.

Understood (2022). *Annual Report*. New York: Understood Inc. Retrieved from: https://annualreport.understood.org/

Vartanian, O., Navarrete, G., Palumbo, L. & Chatterjee, A. (2021) Individual differences in preference for architectural interiors. *Journal of Environmental Psychology, 77*, 101668.

Vitoratou, S. et al. (2023) Misophonia in the UK: Prevalence and norms from the S-Five in a UK representative sample. *PLoS One, 18*(3), e0282777.

Willis, K. & Cross, E. (2022) Investigating the potential of EDA data from biometric wearables to inform inclusive design of the built environment. *Emotion, Space and Society, 45*, 100906.

12 Spaces for collaboration

Editors' introduction

Many organisations believe that their future success is dependent on effective collaboration. So, there is much focus on designing workspaces to support collaboration but maybe less so on designing workplaces for concentration, as reviewed in the next chapter. Nevertheless, collaborative work also requires bouts of working alone, e.g. meeting preparation and development of shared ideas, so those designing for collaboration cannot ignore the challenges of designing for solo, focused, work.

Interaction is a precursor to collaboration and takes place in multiple settings, both in-person and online, with the latter supported by new technology and applications. Regardless, research illustrates several ways that workplace design can support collaboration (in addition to providing the tools required for the core activities). For example, Nicolai et al. (2016) found that using a variety of rooms in different locations throughout the workplace can enhance group creativity or facilitate a process transition from one project phase to another. They also learned that creativity was encouraged when teams transformed spaces to "make them their own",

In activity-based working environments, teammates may have less daily exposure to each other, so Appel-Meulenbroek, Groenen and Janssen (2011) recommend providing team spaces to support intra-team interaction and to promote trust among teammates who otherwise might be dispersed throughout the building. However, some teams do not welcome members of other teams using a meeting space they perceive that they "own", i.e. in the middle of their team zone (van Meel, Martens and van Ree, 2010).

DOI: 10.1201/9781003390848-12

Researcher perspective

Nigel Oseland

Lack of research

Heerwagen et al. (2004) point out that "given the high interest in the topic of collaboration, there is a surprising dearth of research on the link between collaborative work processes and space". While much research has been carried out on the psychology of interaction, interaction is not quite the same as collaboration and the implications of the research therefore slightly differ. Over the years, guidance has emerged on how to design collaboration spaces. However, it is mostly based on case studies, i.e. feedback from best practice workplaces, rather than empirical research. As Hua (2010) commented that "the effects of the physical environment on collaboration at work tend to be overlooked". Considering the increased level of on-line interaction post-pandemic, there appears to be relatively little research exploring the differences between in-person and on-line collaboration.

Defining collaboration

To improve the design of collaboration spaces we first need to understand what is actually meant by collaboration. Marinez-Moyano (2006) captures the common interpretation of collaboration as "a recursive process where two or more people or organizations work together to realize shared goals". Indeed "if tasks are not interdependent then there is no need nor reason to collaborate. Individuals working alone can do the work" (Cohen and Mankin, 1998).

While the notion of interdependency is key to collaboration, many experts in the field believe that true collaboration is about creating something new beyond the capability of the individual collaborators. For example, Cohen and Mankin (1998) suggest that "collaboration enables parties to reach a synthesis i.e. a new conclusion or idea that incorporates the insights of each party but goes beyond each". Similarly, Schrage (1998) argues that collaboration is an "act of shared creation" which involves "two or more individuals with complementary skills interacting to create a shared understanding that none had previously possessed or who could have come to on their own".

Schrage (1998) proposed that most organisations do not have the conditions in place to support people working together to achieve a common goal and therefore diluted notions of teamwork often mask genuine attempts at collaboration. The need for trust to foster collaboration is a well-documented basic condition. Cohen and Mankin (1998) note that "collaboration involves personal relationships between people ... it involves willingness to trust someone enough to work through a conflict". Similarly, Jarvenpaa, Knoll and Leidner (1998) note that "collaboration is a social process and trust is an important contextual factor for both collaborative and virtual relationships". In their review of workplace friendships, Hamilton, Virha and Almedia (2023) report many advantages of social ties at work including building trust and enhancing collaborative reciprocity. Building trust through socialising is important for nurturing collaboration. Management guru Tom Peters (1992) once commented "while we fret ceaselessly about facilities issues such as office square footage allotted to various ranks, we all but ignore the key strategic issue – the parameters of intermingling". Therefore, while collaboration is more complicated than interaction per se, interaction (particularly social) helps build trust and is therefore a prerequisite for true collaboration.

Psychology of interaction

Much of the environmental psychology research has focused on the perception of spaces and how they affect interaction. Notably, Irwin Altman (1975) brought the various theories of space together into one unifying theory, and his version of privacy is akin to levels of preferred interaction. Control over the level of required interaction will impact the likelihood of true collaboration. "Not achieving the desired level of privacy will result in discomfort and stress with too little privacy leading to feelings of overcrowding and too much privacy creating social isolation". Furthermore, Altman proposed that personality factors affect perceived privacy, and I found similar findings in my own research, see Chapter 10 and Oseland (2013). Spaces for interaction and collaboration will require different levels of privacy to cater for different personality types. For example, introverts prefer to interact and collaborate in smaller groups and more private spaces rather than speak out in large formal meetings.

Osmond (1957) introduced the term "sociopetal" space, a space designed to foster interaction, and "sociofugal" space, which discourages interaction. This categorisation applies to the layout of buildings where sociopetal layouts stimulate interaction as routes merge and overlap but, in contrast, buildings with lots of enclosed space, corridors and little common space may be considered sociofugal. Another basic application of Osmond's theory is how meeting spaces are designed and arranged for different types of interaction. For example, breakout spaces that do not offer some level of privacy, drinks, comfortable seating or a pleasant design are sociofugal and will discourage interaction. Seating arrangements also appear to influence the interaction patterns of the group, for example participants of a group generally welcome others into the group by repositioning themselves to form a circle.

Online versus in-person collaboration

Cohen and Mankin (1998) point out that the effectiveness of an organisation is based on how quickly teams can come together to respond to changing business needs and if this physical coming together cannot be achieved quickly, then it will need to happen virtually. Before the Covid-19 pandemic, research into effectiveness of online/virtual collaboration was important due to the increasing globalisation of organisations which resulted in some team members being dispersed. Of course, the effectiveness of online collaboration is even more important post-pandemic as most teams will now have people working remotely, some or all of the time.

Studies comparing the performance of virtual and physically co-located teams found that virtual teams tend to be more task-oriented and exchange less social information than co-located ones (Chidambaram, 1996; Walther and Burgoon, 1992). The researchers suggested this would slow the development of relationships, which is important as strong relational links have been shown to enhance creativity and motivation. Other studies conclude that face-to-face team meetings are usually more effective and satisfying than virtual ones, but nevertheless, virtual teams can be as effective if given sufficient time to develop strong group relationships (Chidambaram, 1996). This research implies the importance of facilitating social interaction in the workplace, and between team members (virtual and co-located) when the team is initially forming. Hua (2010) proposes that repeated encounters, even without conversation, help to promote the awareness of co-workers and to foster office relationships.

For team interactions, the online meeting platforms, such as Teams and Zoom, are a considerable improvement over teleconference calls where those on the call have no sight of others in the meeting. However, Nova (2005) points out that physical proximity allows the use of non-verbal communication including different paralinguistic and non-verbal signs, precise timing of cues, coordination of turn-taking or the repair of misunderstandings. Psychologists note that deictic references are used when meeting in-person on a regular basis, which refers to pointing, looking, touching or gesturing. Newlands et al. (2002) analysed interactions of two groups performing a joint task either in-person or using a video-conference system. They found that deictic hand gesture occurred five times more frequently in the in-person meeting compared to the virtual interaction. Barbour and Koneya (1976) famously claimed that 55% of communication is non-verbal communication, 38% is done by tone of voice, and only 7% is related to the words and content. Clearly, non-verbal communication is a key component of interaction and virtual interaction systems need to facilitate this basic need, especially in the early stages of team forming.

However, the research cited above was all carried out pre-pandemic when interaction on-line was a novelty, and the supporting technology was undeveloped and less reliable. Many workers have adapted to interacting on-line and it has some advantages, beyond the obvious convenience of not having to travel to meetings (both overseas and local). In some of the recent interviews I have conducted, I heard that on-line meetings can enhance interaction and sometimes enable trust to form more quickly. For example, on-line meetings mean that all participants can be more easily seen (assuming they all fit on one screen) than if sitting around a large rectangular table. Similarly, the participants may have "level pegging" as the screen offers a more equitable layout than, say, meeting spaces where the participants vie to be nearer the chairperson. On-line meetings may offer more control, such as being able to control the volume of speakers. Nonethless more research is required on the pros and cons of true collaboration taking place on-line rather than in-person.

Design and location of collaboration spaces

The physical co-location of team desks also facilitates collaboration. A seminal piece of research carried out by Allen (1977) demonstrated that the probability of two colleagues interacting is inversely proportional to the distance separating them, and it is close to zero after 30 metres. In their comprehensive review of the social science literature, Fayard and Weeks (2005) stressed the importance of providing a range of spaces for collaboration that are close to the team. These spaces do not all need to be dedicated collaboration spaces but can be other legitimate and accessible spaces for interaction, such as service and amenity spaces.

Brager et al. (2000) suggest that innovation thrives on all sorts of interaction so we should:

> increase opportunities for spontaneous encounters ('casual collisions') through the use of internal 'streets' and 'neighbourhoods' with cafes and coffee bars.

Good interaction spaces should offer the appropriate level of privacy, which will depend on the content of the interaction and the personality. Interaction spaces do not need to be totally enclosed, but in addition to open and public breakout spaces and cafes, etc., we need to provide semi-hidden spaces that are slightly remote from the main team area e.g. "nooks and crannies".

Green (2012) reminds us that behavioural norms are as important as the design of the spaces: "people must feel they have permission to linger in informal collaborative areas and that comes from watching how other people, especially managers and executives, use or ignore those areas". Fayard and Weeks (2005) revealed the following conditions for creating successful interaction and collaboration spaces.

* *Proximity* – as the frequency of all forms of communication decreases over distance, the proximity of spaces for interaction is of utmost importance.
* *Accessibility* – ease of accessibility and the known availability of spaces for interaction is key, they need to be conveniently located with appropriate visual access and easily located.
* *Privacy* – interaction spaces should provide a sense of perceived visual and acoustic privacy, which does not necessarily mean that full enclosure is required.
* *Legitimacy* – people need a valid reason for being in the space where interactions may take place, e.g. a copy/print area or stairwell/corridor.
* *Functionality* – the layout of the furniture, equipment provided, environmental conditions, amenities and capacity all impact the suitability for different types of interaction.

Brager et al. (2000) argue that "teams need 'team spaces' because team members need to meet frequently, and often in unplanned sessions, so more space should be devoted to group work areas and group tools and team members be co-located to enhance ease of meeting". They propose that an increase in space devoted to teamwork will decrease reliance on the personal workspace and ultimately lead to the demise of the private office.

Table 12.1 Categorisation of collaboration space

Legitimacy-based categorisation (Hua et al., 2010)	Functionality-based categorisation (Coffman, Smethurst and Kaufman, 1999)
Teamwork-related include conference rooms, formal settings, open meeting areas and team rooms in which groups have priority.	**War rooms** represent attempt to improve the collaboration between people and real-time information.
Service-related spaces refer to shared service areas in which copiers, printers and other shared office equipment are located.	**Creativity centres** where play, visualisation and out-of-the-box activities create lateral shifts in thinking.
Amenity-related spaces include kitchens, coffee areas and lounges.	**Collaboration centres** hold the middle ground, a balance of the need for creative thinking and access to real-time information.

Hua et al. (2010) identified three categories of collaboration space, each of which offer a high level of legitimacy and varying degrees of functionality, see the table below. In an earlier study, Coffman, Smethurst and Kaufman (1999) categorised collaboration spaces according to their functionality, particularly how they support sharing real time information or facilitating creativity (Table 12.1).

Hua et al. (2010) conducted original research on the preferences for collaboration space and, in particular, the location of such spaces in the office. They recommend that uniformly distributed clusters of shared spaces, or local hubs, are provided rather than banks of centrally adjacent spaces. However, in practice, central banks of collaboration spaces are likely to be better managed. A balance is required of distributed nodes for spontaneous interaction, local hubs for team collaboration and central facilities for planned (client) presentations and training.

When bringing colleagues together, the reason for the interaction seems to be a more logical starting point rather than, typically, first considering the location and size of space required. A meeting room is only one possible option for facilitating an interaction and, depending on the reason for the interaction, not necessarily the best environment. The best space, its design, layout and setting, is dependent on the purpose of the interaction.

The main purpose of interactions at work, all fundamental to collaboration, needs to be recognised: sharing information, making decisions, resolving problems, generating ideas and socialising (Oseland et al., 2011). Preferred environments suited for different interactions are summarised as follows.

- *Sharing information* – new and complex information needs explaining by the creator either in a local meeting room, with good projection facilities, or by webinar when the recipients are geographically distributed; the information can be sent out by email in advance but the recipients should be given the opportunity to respond in a shared forum rather than through numerous emails.

- *Making decisions* – although some decisions involve a large number of stake-holders, in general decisions are made more quickly within smaller groups; consider locations which minimise interruption and keep the focus of the group, for example a discreet meeting room or off-site conference room.
- *Resolving problems* – the office, especially if on view, is not always the best place to resolve personnel problems; consider a quiet cafe or restaurant, the key is to not be overlooked by colleagues.
- *Generating ideas* – creativity, brainstorming and flow of ideas can benefit from taking place outside of a formal setting; consider different and stimulating spaces and ensure there is good equipment for capturing ideas and space to break out into smaller groups if required.
- *Socialising* – spaces offering food and drink, recreating the "watering hole" are best; also consider meeting outside of the office building to clarify the break from work.

In my "Landscaped Office" concept (Oseland, 2022), in addition to formal meeting rooms for collaboration, I propose the following spaces.

- *Touchdown space* – Typically a large desk or table, sometimes referred to as a "kitchen table", provided for intermittent laptop work or to facilitate visitors or groups of cross-functional colleagues occasionally working together.
- *Project room* – This is a cross between a meeting room and brainstorm area that may be allocated to a team for the duration of their project.
- *Brainstorm area* – Usually a meeting room with additional whiteboards, white-walls or smart boards. However, such spaces do not need to be fully enclosed if located slightly away from the main desks.
- *Informal meeting areas* – A small table and set of chairs in the open plan support impromptu informal meetings. Informal meeting areas may comprise soft seating and sofas for a more informal setting.
- *Multimedia hub* – These are usually furniture solutions, such as Steelcase's "media:scape", that facilitate collaboration in the open plan.
- *Banquette seating* – Usually a rectangular or circular booth used to create an informal semiprivate meeting space. This may be located in a breakout space or staff café/restaurant and used outside of lunchtime.
- *Breakout space* – An area allowing the occupants to break away from the main desk area. Breakout spaces may facilitate informal meetings and social interaction, are usually located near a small kitchen or vending area and include a range of furniture styles.

Conclusion

True collaboration is when two or more people come together and create something that they could not produce alone. Clearly team interaction and collaboration are important in many industries and can lead to new innovative products, cross-selling and faster time to market. Trust is a prerequisite of collaboration and trust can be built through social interaction. A mix of spaces is required that support different interactions and are designed to encourage their use by a range of people. Mingling spaces, interaction nodes, breakout spaces and games areas etc. will all contribute to fostering a collaborative working environment.

Practitioner perspective

Amy Manley and Ellen Keable

Introduction

Work is a fundamentally social process, often involving interdisciplinary interaction and team-based problem-solving. In an increasingly dispersed and asynchronous world, time and place for being together are more important than ever. Workplaces serve collaboration – to share and create in ways we cannot do alone.

Well-designed workplaces advance collaboration by promoting awareness of who people are, what they know and what they are doing. Social networks and collegial trust benefit from face-to-face access to places to interact, meet and work together easily with control for privacy.

Our clients value the effects of places on collaboration, performance, innovation and wellbeing. We work with them to define:

- purpose of place for their culture and people,
- specific behaviours and work processes involved in collaborative work for different job roles,
- design requirements.

We then leverage neuroscience, design experience and evaluation to select and combine the right spatial, design and environmental characteristics that promote interaction, build community and trust, facilitate engagement, knowledge sharing and team building.

Since individual work in many professional organisations can be done anywhere and anytime, coming together face to face, i.e. "collaboration" is an essential purpose of place. However, collaboration involves multiple work behaviours and is more complex than providing openness and a variety of workspaces. Simply providing open plan workspaces with a few options for group meetings will not support collaboration. In fact, poorly designed open plan can suppress interaction through a lack of privacy and concern about not to disturb others (Bernstein and Turban, 2018).

Multiple behaviours and needs for different spaces

Collaboration draws on diversity of expertise, experience and perspectives for effective learning and innovation. Getting to know what others know and the work they are doing, building trust and comfort for sharing skills and ideas and learning how to work together comes from awareness of each other, informal interaction, social networks and shared tasks.

Access to the diversity of knowledge and organisational resources is an important aspect of collaboration. One barrier to interdisciplinary collaboration is affinity bias – that people tend to communicate most effectively with people who are like themselves, which is compounded when they are grouped together (Brill, Keable and Springer, 1996). This psychological force tends to homogenise contacts, suppressing different perspectives. Affinity for similarity also tends to pull together and isolate special interest groups. When we do not take active steps to encourage interaction and collaboration, we reduce diversity's creative benefits.

Face-to-face collaboration involves various levels of access, types of interactions and space needs.

- *Awareness by seeing and hearing* activities, people, groups, resources and events. Awareness improves our ability to know others so we can coordinate, share information and offer resources and assistance. We design to improve awareness in public and semi-public zones with visual connections, group identity display, space for events and exhibitions in high traffic zones, hallway conversation niches and space planning to promote chance encounters while protecting individual and group private work areas.
- *Brief face-to-face interactions* (accidental or intentional) to get to know each other, ask questions and follow up on previous meetings or emails are useful adjuncts to more formal interaction. Space planning combines destinations where people can see each other and briefly chat nearby (standing and sitting options) where they do not disturb concentration in individual or group workspaces.
- *Meetings* (scheduled or unplanned) and *Extended Group Work Sessions* with two or more people often involving visual reference materials, virtual participants and other tools or displays used by the organisations and their disciplines We design meeting spaces to incorporate the technology and tools for virtual and physical collaboration, as flexibility for multiple types of discussions and teamwork. Places for individual work nearby are important for a place to do individual work in between meetings when meetings are the primary reason for coming to the workplace.
- *Toggling mix of individual work and side-by-side discussions* is a frequent collaboration mode for project teams in scientific and creative disciplines. These are sometimes designed among individual workspaces for highly interactive groups as well as separate shared team spaces to reduce distractions to solo concentrated work.

The proportion of individual work requiring concentration, virtual collaboration and face-to-face collaboration in the workplace varies by organisational types, global versus local teams, remote work policies, job roles, disciplines, personalities and project cycles. Understanding and design for differences in flow, cycles and styles require variety and flexibility in setting types.

For example, a client's engineering product development teams had more meetings and informal interaction for project definition and resourcing at the beginning of their cycles, more toggling between individual and one-on-one touch point discussions during development and extended group work sessions at project milestones for review and completion. This variability in work required collaboration spaces ranging from spaces for intense focus, flexibility to bring teams together for several months at a time in project "war" rooms as repository for thinking and visualisation of ideas, to many proximate, unreservable places for quick connect idea sharing and informal interaction.

In contrast, we have clients who have constant need for either back-and-forth interaction or problem-solving, and other clients whose need for focused concentration preclude places for interaction near individual space.

Working alone is still important!

Creating opportunities for interaction needs to be balanced with continued support for individual concentration. Workplace design needs to enable an easy flow between solo and interactive work to maximise useful interactions without disturbing others' concentration. Important design qualities are natural gathering places within and between group areas, visual access to co-workers to know who's available, discussion books in high traffic areas and protected meeting space and project rooms for focused group work.

The process of design should also be collaborative

Designing for collaboration is complex as it needs to address the cultural and behavioural aspects of change as well as space design. The process of design requires collaboration to define and execute the purpose of place through design, technology, policy and training.

Work environments designed for "collaboration" are often unsuccessful for reasons having little to do with design and more to do with understanding and support for expectations of changes in work processes, technology and organisational change. Technology problems and a lack of understanding of how to use it well can also impede virtual and physical co-presence in meeting spaces. Managers and employees who feel left out of the design process can feel misunderstood and have unmet expectations and needs for design, technology and/or policies that affect their experience.

Designing for interaction and collaboration

Our research about liveable density (Keable, Manley and McGregor, 2018) identified an essential design attribute – choice and control – over how, where and when we interact and share space. We showed that increasing workplace density without providing zoning and destinations, adequate control and workspace options leads to a sense of crowding, with negative impacts on employee performance, engagement and wellbeing. Increasing density can also hurt collaboration! When people feel crowded, they often withdraw and send "don't bother me" body language, avoid eye contact, wear headphones, even exhibit hostility.

The power of control over one's work environment allows individuals and teams to self-select and move fluidly to and through those spaces that best support their work and feel good doing so. Our research and experience show the positive impacts on collaboration and team performance from working in spaces that provide the right lighting, acoustics, limit interruptions and give a sense of safety, privacy and choice.

As strategists, designers and architects, we have the tools to design great spaces for group work, interaction and building trust. This is what is most magical about design – fulfilling what appear to be opposite needs. Understanding the ecosystem of human factors (and behaviours) and spatial attributes allows us to make the right business and design decisions to balance comfort, privacy and interaction. In today's current work environments that are focused on connection, we need people to feel empowered and confident to use all spaces to support their work.

Design for collaboration has traditionally focused on enclosed environments like meeting rooms with tables, chairs, display, and technology, and open collaboration pop-up areas with little technology or a sense of privacy.

Increased openness and density are often used to promote communication and collaboration by bringing people closer together with fewer visual and physical barriers. Open collaborative space is frequently used to increase awareness and access. But lack of visual or physical barriers increases a heightened sense of exposure and prevents building trust from sharing insights, work in progress and candid conversations that are contrary to effective collaboration. We also know individuals all work differently. Work environments need to provide a range of workspace types for variety and choice. That immediate need for connection, sharing of ideas and learning about each other is often best accommodated by easily accessible, proximate, visually connected and unscheduled spaces.

Design elements in practice

There are better and more supportive ways to connect people and support collaboration when attention is paid to human needs and environmental factors. As an example, a recent workplace project for Campbell Soup Company's headquarters in Camden, New Jersey supports a range of collaboration behaviours and human needs indicated by the spaces regular use and demand for the new open and collaborative spaces.

Early in the strategic discovery process, Campbell's management defined their organisational goals and success factors for financial, operational, cultural and human success. With a focus on building on the momentum of Campbell's business creating people-to-people energy, defined project goals include:

- consolidating several offices into its HQ campus to build culture and improve connectivity, innovation, delivery and brand connection,
- improving employee experience while increasing headcount and density,
- connecting people across the organisation to promote multidisciplinary innovation,

- creating a modern, welcoming, energetic and connected work environment that responds to the variety of people's work needs and team innovation,
- defining a highly responsive, well-researched, hybrid work policy specific to their work and culture.

We responded to leader visions and insights with workplace strategy and design that transformed how people came together in the new hybrid working model. The previous workplace strategy was last implemented 15 years ago which followed the design trends at the time with a universal planning concept and fewer types of collaborative spaces. The new workplace strategy responds not only to a new way of working but also the goal of creating more energy and team connections. With that in mind, the design approach celebrated elevated brand neighbourhood concepts for the two business divisions as well as identity for the corporate functions. Understanding that each Division and Corporate Function works and collaborates differently throughout the day and within project work cycles, we aimed to find the right mix and quantity of work settings in size, degree of openness and enclosure. We designed their workflows of individual focus, interactions and intense group work. Responding to unique individual and team requirements, the palette of places included individual workstations, enclosed offices, focus rooms and impromptu places for touchdown (Figure 12.1). Meeting places provided different configurations and posture changes in huddle rooms, enclosed meeting and flexible, dynamic group project spaces and rooms.

Addressing goals to increase awareness of and interaction between multidisciplinary groups, we distributed team connection areas at key intersections. We designed for human comfort – using varying levels of screening to mitigate auditory and visual distractions which gave a sense of enclosure for comfort and conversation privacy.

Design solutions include built, enclosed "Pods", hanging, custom branded felt panels, plantings/biophilia and bookcase/display dividers. Colour, materials, pattern, lighting levels and types vary throughout to change the look and feel of the team connection points (Figure 12.2).

Spaces are technology enabled with fixed and mobile monitors and portable power. There are varying degrees of openness from fully enclosed pods, semi-enclosed booths, to fully open (no walls but screened). Screening distance from the individual work settings was deliberate to provide barriers, set boundaries and limit distractions (Figure 12.3). The work settings had varying furniture solutions from stand-up meeting tables for change in posture, small, seated height meeting configurations, lounge settings for small and large groups and spaces with multiple posture and furniture options for large group workshops.

Positive feedback from the users of these highly used and valued spaces Illustrates the following principles when designing collaborative environments.

- *Create community* – Provide a sense of connection to the organisation, team and each other by celebrating brand and providing easily accessible places to connect with the right proximity to individual work settings. (Think of the three bears, "not too close, not too far, but just right").

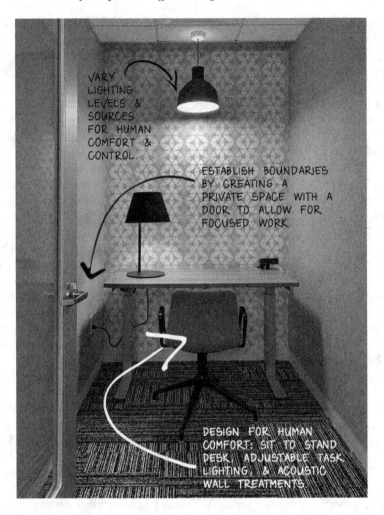

Figure 12.1 Focus room.

- *Establish social norms* – Understanding the why of the design is the basis for understanding how to use and share space.
- *Create inviting destinations* – Use zoning and design to distinguish different types of places for formal meetings, collaboration and focused work. Make collaboration spaces inviting with attractive, comfortable and functional design.
- *Amplify spatial variety and choice* – One size or type does not fit all. We all work differently as individuals and teams depending on what we are doing, our roles and personalities. Support workflows by providing places to move from one type of collaboration to the next seamlessly.
- *Celebrate visual variety and create focal points* – Deliberate changes in lighting levels, materiality, furnishings, levels of acoustics, orientation of spaces create positive spatial end-user experiences.

Figure 12.2 Group-specific collaborative space.

Figure 12.3 Screened collaborative space.

- *Establish physical, visual barriers and boundaries* – This is a big one! An open collaborative space may not want to be totally open. As humans, we gravitate to places that make us feel secure, safe, cocooned, and have a sense of separateness from other groups or people yet with choices for when we want to be with others. Spatial characteristics of scale and proportion are essential. Again, the size needs to be just right. Having the ability to regulate visual, acoustic and physical access and information flow gets back to our need for control.
- *Design for human comfort* – Human comfort is more than posture and ergonomics. We must consider acoustical materials, lighting, ceiling enhancements, nature views, biophilia and thermal comfort.
- *Design for flexibility* – Modular, scalable solutions that adapt and change with people and work needs overtime extend the usefulness and longevity of space.
- *Enable technology and tools* – We live in a digital world requiring portable, pervasive technology. Being connected and able to share content is expected in most spaces today. But training is essential to optimise use and be realistic about organisational resources. Overpromising and underdelivering on technology is most often the norm.

Conclusion: Purposeful design for collaboration and connection

Leveraging design's effects on collaboration, performance, innovation and wellbeing starts with understanding the unique purpose of place and work behaviours for each organisation. Having defined those, design professionals have the tools and experience to create high-performing work environments if there is focus and intention in designing for human-centric needs and requirements.

Great collaborative workplaces require more than variety and choice of spaces. They need to mirror the fluid work behaviours and spectrum of human needs for privacy, connection, and wellbeing. The magic of design is that it can satisfy what seems like opposing solutions by leveraging the following essential design elements.

- *Physical characteristics of the space* – Scale, proportion, ceiling height, windows, views and materials.
- *Work settings with a variety of space types* – To give people more choice, control and flexibility in where and how they work. Include features people can use to regulate their visual, acoustic and physical access and information flow.
- *Zoning* – To separate activity levels, establish group territories, and enable appropriate soundscapes for focused and interactive work.
- *Attractive destinations* – Designed for different types of work.
- *Barriers and boundaries* – To reinforce people's sense of control over their space and send messages about appropriate uses.
- Visual variety and focal points – With deliberate changes in orientation and circulation to create positive spatial experiences.
- *Acoustics and lighting* – Which affect focused concentration, awareness and collaboration, perceptions of size, zoning and territory.

- *Choices about where and how we work to control our own experience* – Choices can be provided with access to various places in the office and opportunities to work off-site and at home.
- *Technology and leadership support for mobility* – Work setting variety and choice only work if people are enabled to use them.

References

Allen, T. J. (1977). Managing the Flow of Technology: Technology Transfer and the Dissemination of Technical Information within the R&D Organization. Cambridge: MIT Press.
Altman, I. (1975). *The Environment and Social Behavior*. Monterey: Brooks/Cole.
Appel-Meulenbroek, R., Groenen, P. & Janssen, I. (2011). An end-user's perspective on activity-based office concepts. *Journal of Corporate Real Estate, 13*(2), 122–135.
Barbour, A. & Koneya, M. (1976). *Louder than Words: Nonverbal Communication*. Columbus: Merrill Publishing Company.
Bernstein, E. & Turban, S. (2018). The impact of the "open" workspace on human collaboration. *Philosophical Transactions of the Royal Society B 37320170239*.
Brager, G., Heerwagen, J., Buaman, F., Huizenga, C., Powell, K., Ruland, A. & Ring, E. (2000). *Team Spaces and Collaboration: Links to the Physical Environment, Final Internal Report*. Centre for the Built Environment.
Brill, M., Keable, E. & Springer, T. (1996). Research on Using Design to Support Dynamic Change and Robust Stability. BOSTI Associates.
Chidambaram, L. (1996). Relational development in computer-supported groups. *MIS Quarterly, 20*(2), 143–163.
Coffman, B. S., Smethurst, J. B. & Kaufman, M. (1999). Seven Basic Concepts of Design for Creating Collaborative Spaces. Sente Corporation.
Cohen, S. G. & Mankin, D. (1998). *Collaboration in the Virtual Organisation*, CEO Publication G 98-28 (356). University of Southern California: Centre for Effective Organisations.
Fayard, A. & Weeks, J. (2005). Photocopiers and water-coolers: The affordances of interaction. INSEAD Working Paper Series. Fontaineblau.
Green, B. (2012). *What It Takes to Collaborate, Research Summary*. Zeeland: Herman Miller Inc.
Hamilton, O. S., Virha, J. & Almedia, T. (2023). Workplace friendships. *The Psychologist, December*, 39–40.
Heerwagen, J. H., Kampschroer, K., Powell, K. M. & Loftness, V. (2004). Collaborative knowledge work environments. *Building Research & Information, 32*(6), 510–528.
Hua, Y. (2010). A model of workplace environment satisfaction, collaboration experience, and perceived collaboration effectiveness: A survey instrument. *International Journal of Facility Management, 1*(2), 1–21.
Hua, Y., Loftness, V., Kraut, R. & Powell, K. M. (2010). Workplace collaborative space layout typology and occupant. *Environment and Planning B: Planning and Design, 37*, 429–448.
Jarvenpaa, S. L., Knoll, K. & Leidner, D. E. (1998). Is anybody out there? Antecedents of trust in global virtual teams. *Journal of Management Information Systems*, Spring, 29–64.
Keable, E., Manley, A. & McGregor, L. (2018). How dense is too dense?. *Work Design Magazine*. https://www.workdesign.com/2018/01/how-dense-is-too-dense/
Marinez-Moyano, I. J. (2006). Chapter 4: Exploring the dynamics of collaboration in interorganizational settings. In Schuman, S. (Ed.) *Creating a Culture of Collaboration:*

The International Association of Facilitators Handbook, San Franciso: Jossey-Bass, pp. 69–85.

van Meel, J., Martens, Y. & van Ree, H. (2010). *Planning Office Spaces: A Practical Guide for Managers and Designers.* London: Lawrence King Publishing.

Newlands, A., Anderson, A., Thomson, A. & Dickson, N. (2002). Using speech related gestures to aid referential communication in face-to-face and computer-supported collaborative work. *Proceedings of the First congress of the International Society for Gesture Studies,* University of Texas at Austin, June 5–8.

Nicolai, C., Klooker, M., Panayotova, D., Husam, D. & Weinberg, U. (2016). Innovation in creative environments: Understanding and measuring the influence of spatial effects on design thinking-teams. In H. Plattner et al. (eds.), *Design Thinking Research, Understanding Innovation.* Switzerland: Springer International Publishing, pp. 125–139.

Nova, N. (2005). A review of how space affords socio-cognitive processes during collaboration. *Psychology Journal, 3*(2), 118–148.

Oseland, N. A. (2013). Personality and Preferences for Interaction, WPU-OP-03. Workplace Unlimited Occasional Paper.

Oseland, N. A. (2022). *Beyond the Workplace Zoo: Humanising the Office.* Oxon: Routledge.

Oseland, N. A., Marmot, A., Swaffer, F. & Ceneda, S. (2011). Environments for successful interaction. *Facilities, 29* (1/2), 50–62.

Osmond, H. (1957). Function as the basis of psychiatric ward design. *Mental Hospitals (Architectural Supplement), 8,* 23–29.

Peters, T. (1992). *Liberation Management: Necessary Disorganization for the Nanosecond Nineties.* New York: Ballentine.

Schrage, M. (1998). Delivering information services through collaboration. *Bulletin of the American Society for Information Science and Technology, 24*(6), 6–8.

Walther, J. B. & Burgoon, J. K. (1992). Relational communication in computer mediated interaction. *Human Communication Research, 19*(1), 50–88.

13 Supporting focused work

Editors' introduction

Although there tends to be more and more attention paid to fine-tuning workplace design for collaboration, there is still a tremendous amount of focused work that happens there – and there always will be.

While some organisations allow working from home as an option, to facilitate focused work without distractions from colleagues, some office workers do not have sufficient privacy or facilities at home and need to come to the office to concentrate.

Large open plan workspaces, with few facilities such as focus rooms, can degrade performance on tasks requiring concentration. Hoendervanger et al. (2018) conducted an experiment in a virtual reality studio and found that "better performance was achieved when complex tasks were carried out in a closed office setting compared to an open plan setting, whereas the opposite was true for simple tasks." Sailer et al. (2021) determined that

> staff are less likely to rate their workplace environment favourably when they have higher numbers of desks within their own field of vision; and when they are facing away from the room with a relatively larger area behind their back compared to the area surrounding them. Aspects of teamwork that are negatively affected include sharing information with others, as well as team identity and cohesion. Focused work (concentration) and working productively are impacted even more so.

Similarly, Brunia et al. (2016) found that "large open workspaces, accommodating more than approximately 15 people, should be avoided due to concentration and privacy issues"; however, they propose that "Large open spaces can be visually and acoustically subdivided in smaller areas."

To facilitate focused work while also supporting collaboration, activity-based working environments were developed and remain popular with workplace designers and consultants. Activity-based workplaces generally provide people with choice and control over where and when they work. For example, Wohlers et al. (2019) demonstrated that activity-based working environments need areas that

DOI: 10.1201/9781003390848-13

support undisturbed solo work as well as those for collaboration. The Wohlers-led team found that merely having workplaces for undisturbed solo work onsite supports job satisfaction, emotion-based plans to stay employed at an organisation and vitality, for instance. Vitality was defined as liveliness and willingness to make an effort.

Researcher perspective

Michal Matlon

Introduction

This chapter discusses the current state of focused work, its importance in the context of organisations, as well as an individual's life. It discusses the workplace factors that can influence the focus of workers, the different types of distractions, and presents six principles for designing an optimal enabling focus environment in the workplace.

The human mind has many superpowers. There is the power of imagination, the power of empathy, of planning, of rational thought or the power of navigating large networks of relationships. It could be argued, however, that all these powers lose much of their strength when the mind is scattered and unfocused.

The number of stimuli our brains receive is enormous, and only by passively or actively choosing which of them to bring to our attention can we function in this world as we do (McCallum, 2023). But there are many signs that focus is missing from people's lives. As a result, they feel frustrated, stressed and dissatisfied with the results of their work.

From following the discourse on work in the last few years, the impression formed is that collaboration is becoming the most important activity now. And it might very well be true that organisations have tended to create siloes, missing out on benefits of multidisciplinary cooperation as a result. However, although ensuring good collaboration is crucial, it loses significance if individual workers cannot dedicate quality time to reflect, create and turn ideas into tangible outcomes. And people do recognise this necessity. The most important work activity, as rated by employees, is still individual and focused work (Leesman, 2023). On average, individuals spend 35% of their time working alone (Gensler, 2023), a substantial amount of time, which accounts for a significant portion of labour expenses. It could all be wasted if used inefficiently.

If individuals highly regard concentrated work and allocate a substantial amount of their time for it, do they find this requirement fulfilled? Apparently not. Over two-thirds (68%) of individuals assert that their day lacks ample uninterrupted focus time (Microsoft, 2023). Enabling sufficient uninterrupted, high-quality and focused work therefore appears to be a significant need and also a significant opportunity.

Researchers and thinkers have already highlighted that focus is becoming a scarce and highly valuable commodity, necessary for thriving today. For example, Newport (2016) argues that "deep work" is an increasingly vital skill for individuals to thrive in the modern economy. He suggests that the ability to focus fully for extended periods is central to sustaining peak performance. And he states that peak performers are one of the three categories of individuals who will flourish in this fresh new environment.

In a similar fashion, others, like Fried and Hansson (2018), highlight the importance of treating employees' attention as a highly precious resource. They advocate for empowering workers throughout the organisation to work in long, uninterrupted focus sessions and argue that current levels of stress, the need for overtime work and inefficiency often stem from a lack of uninterrupted time.

What influences focus?

What, then, are the factors that influence whether people can focus at work?
The authors of this chapter suggest three main categories:

- concentration skills,
- the quality of a person's internal state,
- the environment.

Concentration is a skill that can be developed over time with practice. It enables individuals to attain deeper levels of concentration by filtering out stimuli and sustaining focus for longer periods. The quality of an individual's internal state can affect the frequency of internal distractions. This can be linked to their ability to manage negative emotions and stray thoughts. However, these two factors are largely beyond the scope of this chapter. Let us instead focus on the third factor – an individual's environment. Let's consider this environment as all the external stimuli that a person receives when trying to concentrate. These stimuli might originate from the built environment, the people nearby, and the devices they use.

The main hindrance to concentration is undoubtedly distractions, which are widely recognised by people. Some 73% of people think there is non-stop competition for their attention from various forms of media (The Centre for Attention Studies, 2022). A survey revealed that operating several communication apps results in a loss of 3–9.6 hours per week, varying with the individual's role, and avoidable meetings last 2.8–3.6 hours weekly (Asana, 2023), while keeping 65% of managers from completing their work and doing deep thinking (Perlow et al., 2017). Of employees surveyed in the UK, US and Australia, 62% consider minimising distractions in the workplace important. However, around 55% of offices are reportedly noisy, and over two-thirds of employees report that such noise has a detrimental effect on their ability to concentrate, be productive and creative (Interface, 2019).

Research suggests that distractions increase error rates, with colleague's conversations being rated among the strongest distractors. (Altmann et al., 2014) Some studies suggest that up to 28% of work time might be spent on interruptions and recovering from them (Spira and Feintuch, 2005). It's also said that people tend to compensate for this interrupted work by working faster and so experience more stress and frustration (Mark et al., 2008). One observational study showed that on average, it took workers about 25 minutes to fully recover from a distraction while working on a task (Mark et al., 2005). At the same time, studies such as The Coding Wargames have revealed that, at least for programmers, the factor that distinguishes top-performing individuals from their less successful counterparts when given a series of tasks is a comfortable, distraction-free workspace with ample personal space (DeMarco and Lister, 1985).

How to enable focus?

Given the value of focus for both individuals and organisations, how can we enable it?

Before addressing the physical environment, it's crucial to note that the previously mentioned external distractions can be largely influenced by organisational leadership and culture. It is imperative that employees are encouraged to take focus time and withdraw without feeling obligated to be accessible to co-workers or respond to messages. They also require appropriate training, such as setting up their devices correctly and mastering concentration skills.

There is also the matter of motivation – are individuals motivated to invest their energy into focus and cultivate their concentration abilities? After all, it's not an easy undertaking and requires investing one's energy. Unfortunately, surveys indicate that this may often not be the case. 37% of British employees deem their jobs meaningless, and 13% are uncertain about their job's meaning, resulting in a total of 50% of people who don't feel a clear purpose at work (YouGov, 2015). At the same time, 72% of workers in Europe are not engaged, with 15% being actively disengaged (Gallup, 2023). It's possible that employees may doubt the necessity of engaging in deep work if they feel their job or their employer doesn't really matter.

It's estimated that organisations spend between 40% and 70% of their operating expenses on labour (Delloite, 2020). Given how ineffectively employees spend their time due to various types of distraction and poorly organised meetings, this calls for significant investment in conditions that enable focus. And creating the right focus spaces in the workplace is one of those conditions.

It should be noted that many people have indeed found the right conditions for concentrated work at home or in third places. This was made possible by organisations embracing hybrid and remote working. However, there are a significant number of people who don't have the right conditions to concentrate at home. This could be because they don't have enough space to create focus spaces, or because they are distracted by other household members and personal tasks. There are also those who wish to maintain a clear separation between their working and living environments for reasons of mental health and wellbeing. These people need to be enabled to focus on the workplace.

Focused work principles

Based on currently available knowledge and the experience of the authors of this chapter, the following six principles for designing and building focus spaces are proposed (Figure 13.1).

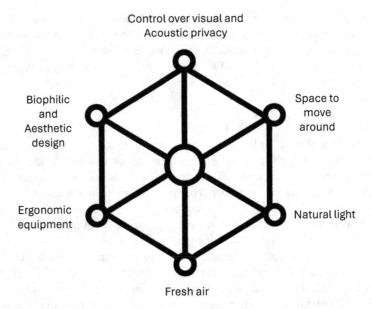

Figure 13.1 Diagram of focus work principles (courtesy of Matlon, 2024).

Full control over visual and acoustic privacy

As demonstrated earlier, distractions are the key factor preventing focused work. This principle gives the worker the ability to fully remove themselves from distractions coming from other people in the workplace.

The simplest and most effective solution meeting this principle is an ordinary, walled-off room, with walls connected to the ceiling, since sound can travel even through small openings in partial enclosures (Brill and Wiedemann, 2001).

Enough space to move around and prevent feeling cramped

In the authors' experience, one of the most common complaints from employees after their organisation has introduced focus booths, which are becoming common, is that they feel claustrophobic and cramped in them.

Focus rooms should allow a person to remain comfortably undistracted for hours at a time. This means being able to stand up, walk around a bit and possibly change postures while remaining in the same room, so that moving around the office and being exposed to different distractions does not disrupt their train of thought.

Natural light

Access to natural light is one of the most basic human needs, and as the amount of time people spend working alone is significant, this need also needs to be met in concentration areas.

Access to natural light helps people maintain their circadian rhythms, improve mood and energy levels, prevent depression and improve general health (International WELL Building Institute, 2020).

Fresh air

This is another basic need that must be met in order to maintain concentration, alertness and mental performance over long periods of time. Elevated CO_2 levels, which in the authors' experience are very common in office environments, have been shown to reduce cognitive performance and have a negative impact on health (Jacobson et al., 2019).

Ergonomic work equipment

An optimal focus space must provide the employee with the ergonomics and comfort of a full-fledged workstation. This means an adjustable, large desk, ergonomic chair, large screen, keyboard, mouse and other accessories to which employees can connect their laptops or where they can log in to remote digital environments (International WELL Building Institute, 2020).

Biophilic and aesthetic design

In addition to the importance of employees feeling comfortable in their work environment to prevent unnecessary negative feelings (which create internal distractions) and to elicit positive ones, it's also been suggested that spaces with biophilic design elements can improve cognitive performance and attention (Browning et al., 2014). If focus spaces are perceived as unpleasant by the employees, it will decrease the probability they will use them and focus well in them.

Although the above principles may seem like common knowledge to experienced workplace professionals, it's important to look closely at how successfully they are being implemented in current workplace design. Now let's take a closer look at the implementation and best practices of focus spaces in the practitioner's part of this chapter.

Practitioner perspective

Cynthia Milota

Planning for focus work

Defining focus in the workplace

Why are dedicated focus workspaces taking a back seat to collaborative spaces?
Does prevailing thinking assume that focus work can be done at the
*   individual workspace or in the open office?*
Or should all focus work be done at home while the office is intended for
*   collaboration?*

The availability of spaces in the office for focus and the permission to use them is at the core of planning for focus work. Focus work is defined as gathering, interpreting and transforming information for business purposes. Focus work also includes solitary reflection, creating new ideas and finding solutions to problems. Unlike routine tasks, focus work requires more thinking and concentration to discover, create or learn something.

Focus work is important as it coexists alongside, while in concert with, collaborative work. The superpowers of the human mind thrive when provided spaces to focus and to collaborate. Key factors to consider in examining focus work from a design and facilities perspective include physical enablers, but also operational and cultural influencers.

The concept of focus work is expanding beyond the individual to incorporate teams of people doing heads-down work *alone-together*. Teams often pursue work of high value to the organisation. The important work of these cross-functional teams is often done in mission or project rooms. Dedicating real estate for both focus and teamwork signals the importance the organisation places on promoting concentration in addition to the ubiquitous collaboration.

Distractions and priorities

Distractions abound, while working from the office or from home. Distractions can be voluntary, such as taking a few minutes to check your email or involuntary, as when someone stops by your desk with a question. Having grown accustomed to quiet places to work from home, spaces for focus work in the office are increasing in demand. Concurrently, there is a de-prioritisation of spaces for focus work in favour of spaces for collaboration.

Hybrid schedules, whereby employees work from the office as well as home, enable more intentionality and segmentation of what types of work are done and where.

This privilege of spaces, permissions and protocols provides employees "the ability to focus without distraction on a cognitively demanding task" (Newport, 2016).

An organisation's protocols and culture implicitly mandate expectations for employee response times to email, text and other communication methods. Urgent, but mundane tasks that do not require much concentration and meetings often occupy the workday. Thus, an assortment of different sorts of work areas is required in a workplace.

Building a workplace for focus work

The experience of working from home during the 2020 pandemic altered the perspective on effective places for focus work. An examination of the development and operations of places which enhance focus work requires consideration of the following factors:

* physical attributes: the spaces and places,
* operational guidelines and social norms: the rules, rituals and protocols,
* psychological influencers: how people approach and accomplish their focus work.

Physical attributes

The physical characteristics for focus spaces outlined earlier in this chapter are re-examined here from a practitioner perspective. For specific examples of how focus work is being conducted in the post-Covid office, a group of Fortune 500 facilities professionals was interviewed using an importance and satisfaction scale.

When examined individually, the physical attributes are interesting, but when considered as a group of criteria for the design of spaces to support focus work, their significance increases. The physical space attributes identified are listed in order of importance for doing focused work well, as assigned by the practitioners interviewed.

Ergonomics

"Sit stands and double monitors are table stakes for focus spaces."

Furniture to support different work postures was indicated as the most significant physical attribute for focus work. With greater attention paid to workplace health and wellness, ergonomics is governed by an organisation's Environment, Health, and Safety (EHS) policies.

Whether one's preference is to sit, stand or walk around while doing focus work, the furniture, technology and lighting must be adjustable to accommodate various body types, postures and work styles.

Fresh air

"It's hard to concentrate when your focus space is hot and stuffy."

Indoor air quality (IAQ) that includes adequate ventilation with filtration and thermal comfort was mentioned as the second most important physical attribute for focus workspaces. Facilities management professionals agree that "hot-cold temperature" issues are one of the most common work order requests.

Individual heating, ventilation, and air conditioning (HVAC) zoning of small spaces such as huddles or focus rooms is preferable, but often cost-prohibitive. Providing adequate air changes and temperature ranges will enhance satisfaction and utilisation of enclosed focus spaces.

Two-way, full acoustic privacy

"My noise cancelling headphones are the only way to get away from the din of the open office."

Open plan office workers are increasingly aware of not only being distracted by office noise, but also being a distraction for those sitting around them. "Hearing, dubbed the 'sentinel of senses' (Banbury et al., 2001) detects and receives information at all times and from all directions. The brain tunes in when speech is recognised; it then diverts attention away from the current task" (Nagy et al., 2016).

While we cannot turn off our ears, people's ability to concentrate in noisy locations is variable. Consider the coffee shop buzz, a distraction for some and a good vibe for others. The rise of the office quiet zone supports those seeking a library-like environment. The integration of sound masking systems engineered to match the frequencies of human speech makes conversations more difficult to hear and understand, providing some level of acoustic separation from others in the open office.

Enough space to move

"Those little phone booths are great for a call, but there's not enough space to work on a project."

The interior industry has responded to the need for focus in the open office by providing furniture-based small portable phone rooms to provide acoustic privacy. While these spaces solve the acoustics issue, they are often too small to serve as a viable solution for focus work. In addition to providing enough space to move around, other criteria for focus supportive spaces include adequate worksurface, lighting and ergonomic features such as furniture adjustability.

Natural light and views to nature

"Daylight and a nice view are a bonus when trying to concentrate, but not as important as quiet and a good desk set-up."

While research has shown that access to natural light is essential for mood, energy and general health, having daylight for focus work was seen as a lesser priority. With many office layouts placing private offices and other collaborative spaces inboard, current interiors are filled with daylight. When it comes to focus work, while daylight is a positive, views outdoors can offer distractions.

Two-way, full visual privacy

"It's so distracting when people walking by make eye contact with me. I need a place with more visual separation to do my focus work."

Visual privacy is linked to distractions, which are hard to manage in an open office environment. In the past, workstations with high panels and even sliding doors solved some of that visual privacy demand. Currently, lower or no panels on open plan workstations are supplemented with small, enclosed focus and huddle rooms. These enclosed spaces with glass walls and doors should include film, shades, drapery or blinds to provide visual privacy. Locating these focus spaces off main circulation aisles or at the end of a space ensures less potential for distraction.

Biophilic and aesthetic design

"The new office has all these cool spaces to work in, away from my desk."

The visual and emotional appeal of the office space tangentially supports focus work, by contributing to the overall employee experience, which equates to satisfaction and ultimately the ability for one to do their best work. Spaces that incorporate biophilia (plants), hospitality (coffee and food), branding (environmental graphics) and cultural artefacts (the stuff of the organisation) are important indicators of employee satisfaction.

Whether the office aesthetic is "resimercial" (combing residential and commercial design aspects), hospitality (akin to a hotel or resort) or corporate, focus spaces may take cues from a living room, a hotel lobby or a well-appointed small conference space (Figure 13.2).

Team rooms

In addition to the physical criteria necessary for individual focus work outlined above, the expanded view of focus work includes teams working *alone-together*. The criteria for the design of team rooms for focus work include:

- *Ergonomic factors* – Sit-stand desks configurable as 4–6 person tables. Adjustable monitors on individual desks are supplemented by wall monitors, mounted high enough for viewing when team members are standing at their desks.
- *IAQ* – Zoning for temperature control in enclosed team space where employees will be working for extended periods of time.
- *Lighting controls* – Ability to control lighting in team rooms, with potential for multiple lighting /options *Visual privacy* – Film on interior glass walls, with the potential for cloaking privacy screens on monitors for confidentiality.
- *Access to daylight* – Unlike individual focus rooms, teams may work in these project rooms for an extended time, making access to daylight and even views outdoors essential. Clerestory windows provide daylight while limiting distractions.

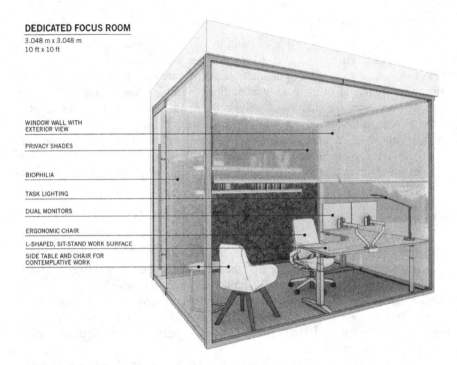

DEDICATED FOCUS ROOM
3.048 m x 3.048 m
10 ft x 10 ft

WINDOW WALL WITH
EXTERIOR VIEW

PRIVACY SHADES

BIOPHILIA

TASK LIGHTING

DUAL MONITORS

ERGONOMIC CHAIR

L-SHAPED, SIT-STAND WORK SURFACE

SIDE TABLE AND CHAIR FOR
CONTEMPLATIVE WORK

Figure 13.2 Illustration of dedicated focus room (courtesy of Ware Malcomb, 2024).

- *Adequate space to move around* – Providing ample width in the room for circulation behind the work seats.
- *Access to focus spaces beyond the desk* – As in open workstation areas, individuals working in team rooms require adjacent, fully enclosed individual focus spaces as well as access to open focus spaces (Figure 13.3).

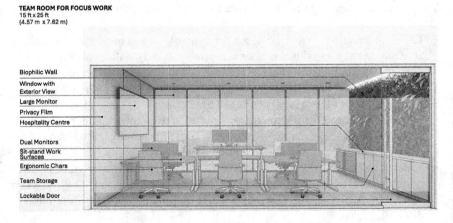

TEAM ROOM FOR FOCUS WORK
15 ft x 25 ft
(4.57 m x 7.62 m)

Biophilic Wall

Window with
Exterior View

Large Monitor

Privacy Film

Hospitality Centre

Dual Monitors

Sit-stand Work
Surfaces

Ergonomic Chars

Team Storage

Lockable Door

Figure 13.3 Illustration of team room for focus work (courtesy of Ware Malcomb, 2024).

Operational guidelines and social norms

Equally important to the physical spaces provided for focus work are the guidelines and social norms to utilise these areas. Aligning the space, with the organisational permissions, is essential for a robust utilisation of focus spaces. Factors for consideration include:

* Cultural rituals that protect concentration and focus:

    Does the organisation, business unit, or manager respect the need for time and places for concentrated work?

* Organisational norms for method and speed of communications:

    What is the implicit response time for communications (emails, texts, etc) and do they take priority over focus work?

* Team agreements, whereby members of a team agree to certain protocols and practices for their group:

    What are the agreed upon team protocols that will protect team members' individual and collective focus work?

* Explicit and cultural permissions for working away from the desk:

    Is there a freedom of place, choice, and flexibility to work in the location best for the work task at hand?

* Spaces are available for reservable and impromptu use:

    Can focus spaces be booked for extended periods of work?

* Training on etiquette and protocols to support work types and styles:

    *Are best practices for maximising focus work integrated into learning and development curriculum?**

Psychological influences

The best-designed space for focus work will be unused, if both organisations and employees do not agree upon focus spaces' use and purpose. Trust is an essential component to hybrid work, but also essential when working away from your desk in a dedicated focus space.

Psychological influences include trusting that employees will select the best location for the work task at hand, and will not be subject to presence bias, defined as being rewarded because one is more visible in the office.

Finding the headspace for concentrated work requires not just the provision of physical privacy … but also the psychological privacy … Territorial and physical elements to aid concentration and privacy tend to be ineffective and irrelevant if the mind is unsettled and unfocused on the job in hand.

(Myerson et al., 2010)

While focus work starts in the head, it unfolds in a physical space.

Conclusion: Getting focus work done in the office

Collaboration and focus work are two sides of the workplace coin. Intentionally, planning work around collaborative and focus work give employees some level of choice and control. However, distractions, interruptions and changing priorities impact the best-laid plans.

Physical, operational and psychological factors influence focus work in the office. Employees desire the ability and permission to work from different locations, supported by various enclosed spaces and quiet zones within the office. Organisationally, there may be implicit and explicit permission to work away from the desk which may limit distractions and interruptions. As importantly, trust conveyed by the organisation, team and manager to their employees proves essential to providing the headspace and the focus space to get work done.

References

Altmann, E. M., Trafton, J. G., & Hambrick, D. Z. (2014). Momentary interruptions can derail the train of thought. *Journal of Experimental Psychology: General, 143*(1), 215–226.

Asana (2023). *The Anatomy of Work: Global Index 2023*. San Francisco: Asana. Retrieved from: https://asana.com/resources/anatomy-of-work

Banbury, Simon P., Macken, W., Tremblay, S. & Jones, D. (2001). Auditory distraction and short-term memory: Phenomena and practical implications. *Human Factors: The Journal of the Human Factors and Ergonomics Society, 43*(1), 12–29.

Brill, M. & Wiedemann, S. (2001). *Disproving Widespread Myths about Workplace Design*. Buffalo: BOSTI Associates.

Browning, W., Ryan, C. & Clancy, J. (2014). *14 Patterns of Biophilic Design*. New York: Terrapin Bright Green.

Brunia, S., de Been, I. & van der Voordt, T. (2016). Accommodating new ways of working: lessons from best practices and worst cases. *Journal of Corporate Real Estate, 18*(1), 30–47.

Delloite (2020). *Human Capital Balance Sheet*. Deloitte United States. Retrieved from: https://www2.deloitte.com/us/en/pages/human-capital/articles/optimizing-the-human-capital-balance-sheet.html

DeMarco, T. & Lister, T. (1985). *Programmer Performance and the Effects of the Workplace*. London, New York, and Aachen: The Atlantic Systems Guild.

Escera, C. & Corral, M. (2007). Role of mismatch negativity and novelty-P3 in involuntary auditory attention. *Journal of Psychophysiology, 21*(3–4), 251–264.

Fried, J. & Hansson, D. H. (2018). *It Doesn't Have to Be Crazy at Work*. New York: HarperCollins.

Gallup (2023). *State of the Global Workplace 2023 Report*. Gallup. Retrieved from: https://www.gallup.com/workplace/349484/state-of-the-global-workplace.aspx

Gensler (2023). *Global Workplace Survey Comparison 2023*. Gensler. Retrieved from: https://www.gensler.com/gri/global-workplace-survey-comparison-2023

Hoendervanger, J., Van Ypersen, N., Mobach, M. & Albers, C. (2018). From activity-based towards needs-based work environments. In S. Nenonen, A. Salmisto & V. Danivska (eds.), *Proceedings of the 1st Transdisciplinary Workplace Research Conference.* Tampere Finland, pp. 12–13.

Interface (2019). *What's That Sound?: The Impact of Office Noise on Workforce Productivity.* Interface. Retrieved from: https://interfaceinc.scene7.com/is/content/InterfaceInc/Interface/Americas/WebsiteContentAssets/Documents/Acoustics%20Survey/wc_am-acousticssurvey

International WELL Building Institute (2020). *WELL v2.* Retrieved from: https://a.storyblok.com/f/52232/x/926ed4f977/well-building-standard-v2-q4-2020-wellapv2-02-03-23.pdf

Jacobson, T. A., Kler, J. S., Hernke, M. T., Braun, R. K., Meyer, K. C. & Funk, W. E. (2019). Direct human health risks of increased atmospheric carbon dioxide. *Nature Sustainability,* 2(8), Article 8.

Leesman (2023). *The Future of Work and the Workplace: Insights from Leesman Global Survey.* Leesman. Retrieved from: https://www.arubanetworks.com/assets/analysts/Leesman_Future-of-Work-and-the-Workplace.pdf

Mark, G., Gonzalez, V. M. & Harris, J. (2005). No task left behind?: Examining the nature of fragmented work. In *Proceedings of the SIGCHI Conference on Human Factors in Computing Systems*, pp. 321–330. New York: SIGCHI.

Mark, G., Gudith, D., & Klocke, U. (2008). The cost of interrupted work: More speed and stress. In *Proceedings of the SIGCHI Conference on Human Factors in Computing Systems*, pp. 107–110. New York: SIGCHI.

McCallum, W. C. (2023). *Attention.* Britannica. Retrieved from: https://www.britannica.com/science/attention

Microsoft (2023). *2023 Work Trend Index: Annual Report.* Retrieved from: https://www.microsoft.com/en-us/worklab/work-trend-index/will-ai-fix-work

Myerson, J., Bichard, J. & Erlich, A. (2010). *New Demographics, New Workspace: Office Design for the Changing Workforce.* Farnham: Gower.

Nagy, G., O'Neill, M., Johnson, B. & Bahr, M. (2016). *Designing for Focus Work.* Retrieved from: https://media.haworth.com/asset/83961/Focus%20White%20Paper_Haworth.Final.pdf

Newport, C. (2016). *Deep Work: Rules for Focused Success in a Distracted World* (First Edition). New York: Grand Central Publishing.

Perlow, L. A., Hadley, C. N., & Eun, E. (2017, July 1). *Stop the Meeting Madness.* Harvard Business Review. Retrieved from: https://hbr.org/2017/07/stop-the-meeting-madness

Sailer, K., Koutsolampros, P. & Pachilova, R. (2021). Differential perceptions of teamwork, focused work and perceived productivity as an effect of desk characteristics within a workplace layout. *PLoS ONE, 16*(4), e0250058.

Spira, J. & Feintuch, J. (2005). *The Cost of Not Paying Attention: How Interruptions Impact Knowledge Worker Productivity.* Basex, Inc. Retrieved from: https://www.interruptions.net/literature/Spira-Basex05.pdf

The Centre for Attention Studies (2022). *Do We Have Your Attention?: How People Focus and Live in the Modern Information Environment.* King's College London, The Centre for Attention Studies. Retrieved from: https://www.kcl.ac.uk/policy-institute/assets/how-people-focus-and-live-in-the-modern-information-environment.pdf

Wohlers, C., Hartner-Tiefenthaler, M. & Hertel, G. (2019). The relation between activity-based work environments and office workers' job attitudes and vitality. *Environment and Behavior, 51*(2), 167–198.

YouGov (2015). *37% of British Workers Think Their Jobs Are Meaningless.* London: YouGov. Retrieved from: https://yougov.co.uk/topics/society/articles-reports/2015/08/12/british-jobs-meaningless

14 Spaces for revitalising

Editors' introduction

Without respite and refreshment, there is a rapid degradation of mood and performance by people at work. When people are mentally tired, they snap at colleagues, do subpar work, have trouble concentrating, and feel stressed (Gifford, 2014). When refreshed, our abilities to get along with each other, to focus, and to process information effectively are restored.

When given an opportunity to step away from their work, people can choose to refresh in a number of ways – some more active (like exercising) and others more relaxed (such as drinking a cup a of tea). Neuroscience research has shown that regardless of how people choose to revitalise themselves, having specific experiences such as seeing or hearing nature will refresh them at work. Introducing revitalising spaces, and these universally refreshing experiences into the workplace, is therefore paramount for a healthy and productive workforce.

DOI: 10.1201/9781003390848-14

Researcher perspective

Angela Loder

Introduction

After the great work-from-home experiment of the pandemic, offices which did not prioritise occupant health and wellbeing are finding it hard to compete with bespoke home offices (Leesman Index, 2020; Marzban et al., 2023), many of which outperform traditional offices in acoustic and visual privacy and work-life flexibility (JLL, 2023; Marzban et al., 2021). This reluctance from employees to do their daily work in spaces that may not meet their needs has renewed academic and practitioner attention on the impacts of the office on employee mental and physical health (Bergefurt et al., 2022; Colenberg, Jylhä and Arkesteijn, 2021; Whelan et al., 2021).

These pressures challenge employers to go beyond risk management to re-imagining what the workplace can be; spaces and places that are regenerative and revitalising, that promote employee resilience and wellbeing, and that echo trends in sustainability that are moving away from minimising harm to rewilding and regeneration (Ejlertsson et al., 2020; Global Green Tag International, 2023; Konietzko, Das and Bocken, 2023). However, gaps remain from the current siloed approach to research, design, and implementation (Herneoja, Markkanen and Juuti, 2022; Jensen and van der Voordt, 2020) which impact an organisation's ability to understand the impact of different health design interventions.

This chapter explores the shift from risk reduction to revitalising spaces for the workplace, current and emerging evidence, tools for implementation and measurement, and alignment with concurrent reporting pressures that can help organisations make the business case for investing in a resilient, revitalised workforce.

Moving from risk-reduction to regeneration and resilience

While the risk-reduction approach continues to be dominant in workplace studies (Bergefurt et al., 2022; Colenberg, Jylhä and Arkesteijn, 2021), additional pressures have begun to expand the original risk-reduction model. First, the recognition that many office workers regularly engage in creative thinking and decision-making versus rote task completion has spurred a more expansive view of productivity. For example, while access to nature has long been associated with better task completion in laboratory studies (Adamson and Thatcher, 2019), there is renewed attention on studies that have linked nature with the creative process (Plambech and van den Bosch, 2015; Williams et al., 2018), or the impact of biodiversity on wellbeing (Carrus et al., 2015; Marselle et al., 2021; Southon et al., 2018).

More complex measures for productivity, such as the CogFx study series that linked poorer decision-making and health with elevated levels of CO_2 (Allen et al., 2016), have also arisen to better reflect the kind of work knowledge workers engage in. Both trends have the potential to shift workplace design to better support the kinds of tasks workers need to do while providing better tools to measure real-world outcomes. Second, the persistent lack of satisfaction from office workers with thermal comfort has given rise to the concept of alliesthesia, which provides compelling evidence that humans are more alert and satisfied with more varied environmental conditions and stimulation (Aghniaey and Lawrence, 2018; Altomonte et al., 2020; de Dear et al., 2020; Dzyuban et al., 2022) as long as they have some degree of control over their environment (He et al., 2022). Allowing for more varied thermal comfort zones and stimuli opens opportunities for spaces to support revitalisation.

Third, there has been a recognition from both the built environment and psychosocial research streams of the impact of health-promoting spaces (Capaldi, Dopko and Zelenski, 2014; Clements-Croome, 2017) and positive organisational culture and structure on worker health, engagement, and energy (Abid, Zahra and Ahmed, 2016; Spreitzer et al., 2005). Examples include the role of daylight in healthy circadian rhythm (Figueiro, 2017; Fonken and Nelson, 2014) and the impact of perceived organisational support (POS) on job satisfaction (Rhoades and Eisenberger, 2002), positive relationships with co-workers and employee thriving (Abid, Zahra and Ahmed, 2016; Dutton and Ragins, 2006). This more holistic recognition of the need to address both built environment and organisational factors offers promise for healthier workplaces. Lastly, the original problems of sick building syndrome arose because of a well-intentioned but siloed approach to energy efficiency that neglected to consider human health. Current research on occupant comfort is once again finding potential conflicts with indoor environmental standards that are based solely on energy efficiency or mean acceptability thresholds (Licina et al., 2021) but also offers opportunities to find holistic solutions that support both human and planetary health (Aghniaey and Lawrence, 2018).

The above research streams point to a shifting approach to what constitutes a "good" working environment. Central to these research streams is the push towards spaces and places that energise, heal, and provide the conditions that enable people to thrive (Bergefurt et al., 2022; Clements-Croome, Turner and Pallaris, 2019). This more salutogenic – or health-promoting approach – views health as a continuum that is influenced by competing pathogenic (disease) stressors and salutogenic resources (Antonovsky, 1987; Roskams et al., 2021). Moving from a risk-reduction to a health-promotion approach starts to shift the workplace from one of health and safety and productivity towards understanding what factors make workplaces a place of healing, flourishing, and vitality. As seen in previous chapters, there is substantial evidence from a wide variety of fields on what kinds of spaces support the health and performance of office workers. But how do they fit into the concept of revitalising, a term more commonly used in community or city-level work? What are common themes and lessons learned that can be combined to create revitalising spaces, and how can practitioners and researchers alike better understand revitalising spaces and their application in the real world? Central to these questions are the role of energy and vitality, rest and recovery, and a holistic approaches and tools that address these polarities.

Vitality – Linking research to practice

Revitalising as a term has more commonly been used to describe community-level activities and impact (Policy Link, 2012), but as applied to workplaces, it reflects a growing awareness of the impact of the workplace on employee energy and vitality (Li, Xu and Fu, 2020; Madsen et al., 2016). Vigour, also known as vitality, is connected to positive health outcomes like reduced anxiety and depression (Keyes, 2002) and is influenced by high levels of emotional and physical energy (Schaufeli and Taris, 2014), all of which benefit the workplace. While the JD-R and related Environmental-Demands-Resources model (ED-R) have provided a framework from which to divide psychosocial and built environment components into either energy-depleting or energy-restorative influences, the revitalising workspace goes further, asking which spaces energise and inspire office workers.

In addition to more obvious resources such as providing freshwater stations and healthy food options (Bucher et al., 2016; Masento et al., 2014), the concepts of alliesthesia and biophilia outlined above reflect a growing awareness that recognising the biological reality of office workers – who respond well to varied environmental conditions throughout the day – leads to better health and performance outcomes. For example, providing access to daylight which shifts through the colour spectrum throughout the day has been linked to increased alertness (Pachito et al., 2018), better sleep (Boubekri et al., 2014), and even a higher salary (Mac-Naughton et al., 2021). Providing different thermal zones and personal control has been shown to increase comfort and productivity (He et al., 2022; Mishra, Loomans and Hensen, 2016), while providing different acoustic zones for different types of work activities helps with task accomplishment as well as providing variety and stimulation (Koukounian and Bourdeau, 2022).

Both well-established and emerging lessons learned from the human response to nature and biophilic design provide insight into design strategies for energy-supporting workplaces. Human fascination with natural scenes are known to draw and hold attention (Berto, Massaccesi and Pasini, 2008; Kaplan, 1993) and be preferred over many human-made scenes (Kaplan and Kaplan, 2005) and can be incorporated via window views, pictures, and artwork (Andreucci et al., 2021; Aristizabal et al., 2021). Surfaces and textures that mimic nature, such as shiny surfaces that reflect like water, or a variety of textures that mimic the variety found in nature, engage more than the visual senses (Kellert and Calabrese, 2015). Nature sounds have the benefit of holding a soft fascination as well as masking conversation (Hongisto et al., 2017), while a variety of natural scents, such as flowers and herbs, can impact memory, emotion, and attention (Pálsdóttir et al., 2021).

More recent studies have shown that human preferences for the fractal patterns inherent in nature are influenced by the scale of visual cues. For example, Abboushi et al. (2019) note that a wider visual frame allows for a higher level of complexity, while design interventions using fractal patterns are most preferred when they shift throughout the day, mimicking the pattern of light through leaves and trees (see Figure 14.1). In practice, this means that views from a window or across an atrium can have more complex natural interventions, while smaller spaces may need simpler designs. Combined with the link between nature and the creative process and the role of biodiversity, making the office more like nature has the potential to keep both the cognitive mind stimulated and the five senses engaged.

Figure 14.1 Biophilic design for an energising view – Daichi Life Insurance, Tokyo (photo credit: Jack Noonan).

Rest and recovery – Linking research to practice

While these design strategies help to enhance vitality, long-standing evidence from organisational management has demonstrated that knowledge work is in itself mentally and emotionally fatiguing (Karasek, 1979; Lund et al., 2005; Nurmi and Pakarinen,

2023), and that the ability to rest and recover is a mediating factor in burnout, sustained performance, and energy (Fritz et al., 2013; Li, Xu and Fu, 2020). This requires different policies, spaces, and places for rest and recovery (Forooraghi, Cobaleda-Cordero and Babapour Chafi, 2021; International WELL Building Institute, 2018).

Wellness rooms that can host a variety of activities, such as meditation, yoga, or rest, should provide a visual and acoustic break for employees. This is particularly important for open-plan offices, given research which indicates a lack of visual and acoustic privacy can engender fatigue (Jacques et al., 2018; Kim and de Dear, 2013; Morrison and Smollan, 2020).

Natural features placed throughout the workplace, including plants, outdoor spaces, and biophilic design have all been shown to help with rest and recovery from taxing mental work (Bergefurt et al., 2022; Colenberg, Jylhä and Arkesteijn, 2021) (see Figure 14.2).

Figure 14.2 Biophilic design and access to nature to encourage walking and breaks – National University of Singapore, SDE4 building (photo credit: Jack Noonan).

Just as important as spaces for rest and recovery are policies and leadership that make employees feel safe in taking it. Health-supportive leadership, combined with design-features to support rest and recovery, can help foster POS and psychological safety, both of which have been linked to positive health and performance outcomes (Herway, 2017; Rhoades and Eisenberger, 2002; Rozovsky, 2015) and which may enable workers to prioritise their wellbeing. For example, recent research on productivity and mental health from a survey of over 3,000 knowledge workers found that those with the best mental health took twice as many breaks from work as those with poor mental health. 61% of those with good mental health reported a high sense of psychological safety, compared with only 10% of those with poor mental health (flow2thrive, 2023).

Leaders play a personal role as well. In a study of leaders in the health and wellbeing industry, those who scored high in their mental and emotional wellbeing reported a greater ability to energise others, maximise the effectiveness and growth of others, and cultivate a positive and collaborative work environment (Wisdom Works Group, 2023). All this challenges traditional views on productivity which measures productivity both by output and by the amount of time in front of a computer.

Lastly, social factors like social cohesion are known impact health and performance outcomes such as stress reduction (Lach et al., 2022), perceptions of a healthy work environment (Lowe, Schellenberg and Shannon, 2003), and individual wellbeing (Kouvonen et al., 2008). One factor that is emerging as increasingly important for diversity, equity, and inclusion is belonging, long known to be linked with positive workplace outcomes such as job satisfaction and commitment and reduced depressive symptoms (Dutcher et al., 2022; Gatt and Jiang, 2020). Lastly, collective wellbeing is emerging as key to understanding how places and organisations impact wellbeing in groups (Morgan et al., 2022). Designing spaces to be welcoming and restorative for diverse needs and abilities is an essential component of creating revitalising workplaces that work for everyone (Lahtinen et al. 2015; Mikus, Grant-Smith and Rieger, 2023).

Tools and measurement

While the above evidence can help organisations understand the link between design and organisational strategies and health and performance outcomes, implementing them to create revitalising workplaces can be more challenging. Existing healthy building certifications such as the WELL Building Standard or Fitwell can help projects implement design and policy practices in a holistic framework to support overall health, while focused tools such as the Biophilic Design Matrix tool (McGee and Marshall-Baker, 2015) can help projects know how to implement nature-based features. If projects are interested in measuring and tracking the impact of their interventions, then survey tools, wearables, and existing performance metrics can be used to track outcomes before and after the intervention.

Surveys that use a more holistic approach to health and wellbeing and that track both organisational and built environment factors are positioned to provide the most compelling, and holistic, evidence on healthy workplaces. For example, the Sustainable Healthy Environments lab out of the University of Melbourne compared 1,400 post-occupancy survey responses from WELL versus other high-performing buildings across Australia, New Zealand, and Hong Kong. They identified key characteristics of high-performing workplaces that spanned both built environment (e.g. layout and interior design) and organisational aspects (e.g. culture and programs) and how certifications like WELL support these workplaces. Not surprisingly, they found higher satisfaction with physical characteristics of the space (e.g., layout, acoustic privacy, and acoustic comfort) in WELL-certified versus non-WELL-certified spaces, but they also found that occupants were more satisfied with organisational aspects such as inclusive culture, engagement, and a positive environment (Marzban et al., 2023). This kind of holistic data is needed to provide the necessary evidence to justify some of the "soft" interventions of a revitalising workplace.

Lastly, organisations are under pressure to report on both sustainability and social factors and often struggle to connect the dots between their built environment and policy interventions for health and wellbeing, their community impact, and their sustainability initiatives (BRAG, 2023; Glazerman, 2019). Using a holistic measurement framework that addresses these different scales of interventions, such as the *12 Competencies for Measuring Health and Wellbeing* (International WELL Building Institute, 2022), can help guide organisations in knowing what to topics to measure across those different scales of impact. Holistic frameworks can also help link building-level benefits, such as employee performance and health, with organisational outcomes, such as increased organisational performance and social impact, thus supporting the business case. This will become increasingly important as pressures to align sustainability, health, and community impact are only expected to increase (European Commission, 2023; Global Green Tag International, 2023).

Moving forward

Revitalising spaces hold promise for workplaces that heal and regenerate and mitigate negative workplace outcomes like burnout. Expanding well-known models on the workplace that balance demands and resources to spaces and policies that energise, inspire, and provide rest and recovery reflect emerging evidence on the need to address the whole person at work, not just the person as a productive unit. These newer models and frameworks are re-thinking the division between health and productivity, positing instead that they are an interrelated ecosystem that influence overall thriving (Butler and Kern, 2016; Wisdom Works Group, 2023). Implementing these practices, however, requires a holistic approach that encourages multiple stakeholders to see how a revitalising workplace can bring both internal benefits, such as improved health and performance, and external benefits, such as better metrics and evidence to show progress on both social and sustainability metrics. Tracking progress over time through existing measurement tools enables an adaptive approach that can encourage flexibility and responsiveness. Lastly, if the overall goal is healthy, resilient workers, workplaces may need to realise that even the best-designed offices are still finding that work-life balance may be better at home (JLL, 2023) and adjust spaces and policies accordingly.

Practitioner perspective

Valerie Jardon and Diane Rogers

Introduction

Work, collaborative or independent, is mentally exhausting. As the pace of modern life continues to accelerate, respite is increasingly essential to counter stimuli overload and the decrease in directed attention that comes with prolonged cognitive exertion (Kaplan, 1995). Without an opportunity for revitalisation, cognitive performance declines as does the effectiveness of an individual's interaction and reaction to others. However, extensive research, as described earlier in the chapter, has demonstrated the restorative psychological and physiological benefits of spaces that are thoughtfully designed for mental and physical refreshment.

With this knowledge, there is a growing organisational interest for the overall design of a facility to provide balance that complements a users' daily activities and anticipates their needs. Thoughtfully designed venues that engage the senses and offer a variety of restorative experiences allow people to naturally replenish their capacity to focus and interact while reducing stress and exhaustion. The following describes an approach to designing for balanced revitalisation.

The importance of variety

To ensure optimum variety, it must be recognised that people revitalise in different ways. Some people refresh in social settings while chatting with co-workers, playing games, or pursuing an activity (*together/we focus*), while others rely on alone time or turn to spaces focused on a concentrated activity other than work (*individual/I focus*). To provide restorative experiences that appeal to everyone, organisations should include a mix of active (*expand*) and thoughtful (*retreat*) zones, enabling users to find the experience that best supports their needs on demand. Figure 14.3 shows the "Balanced Revitalization Chart" and illustrates how these space types are categorised. The scale, placement, and quantity needed for any project will vary based on the site's specific demographics and culture.

As illustrated, each quadrant of the "Balanced Revitalization Chart" represents a different type of revitalising space categorised as Recharge, Connect, Reflect, and Create. Each type offers a form of strategic cognitive rest which is essential for replenishing our neurotransmitters and restoring neurological balance after a state of heightened focus or flow (Fabritius, 2022). Offerings within each space should appeal to the senses, illicit a positive reward and/or emotion, and allow for individual customisation around context and culture; the three basic factors of an individual's aesthetic experience (Magsamen and Ross, 2023).

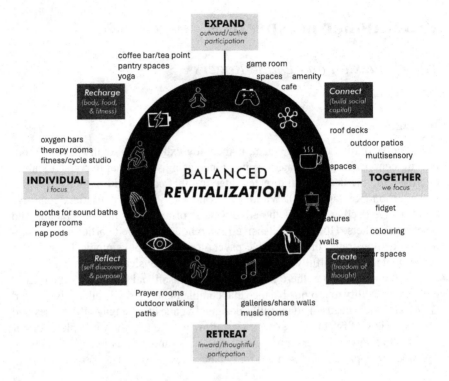

Figure 14.3 "Balanced Revitalization Chart" (graphics courtesy of IA Interior Architects and Lauren Newton).

Space to expand – Energising the individual, connecting the collective

The top half of the "Balanced Revitalization Chart" highlights outward-oriented environments that require a level of active user participation to promote physical revitalisation. On the top left are **Recharge** spaces. These individual (*I-focused*) environments provide users with an opportunity to enjoy a healthy snack or beverage or focus on personal health and fitness. While there may be interaction with others in this space, the primary focus is the user's health and wellbeing. Common space-type examples include coffee, tea, and juice bars, fitness/yoga studios, individual stationary bike rooms, and therapy rooms that offer devices such as electronic massage chairs, handheld massage guns, or compression boot systems that help increase blood circulation. To boost energy and mental clarity the success of these spaces depends on access to natural light, good ventilation, and the inclusion of biophilic patterns that mimic nature in the design (Kellert, 2018).

Connect spaces are group areas for vivacious interaction. They include cafés and lounges that encourage social bonding and benefit from appealing aromas and thoughtful lighting to support interpersonal closeness and communal elements (Crandall et al., 2018). Located in the top right of the chart, these are together (*we-focused*) areas within the floor plan that provide opportunities to build social capital and enjoy a moment of casual fun. They are highly focused on bringing groups of people together to escape the general hustle and bustle of the workplace, but also offer nooks for smaller one-on-one interactions or self-reflection without the user feeling out in the open – an ideal setting for the introverts of the office. Cafés, game rooms, and outdoor areas are popular revitalising spaces in this quadrant.

Seeking a retreat – Sanctuaries of serenity and expression

Inward-oriented environments make up the bottom half of the "Balanced Revitalization Chart". These are spaces that require the thoughtful participation and focus of an individual.

At the bottom left are **Reflect** spaces. These calming environments encourage self-discovery, reflection, and personal wellbeing. They must be serene environments for contemplation that inspire inner focus, incorporating dimmable/tuneable lighting and calming sensory features like water elements, plants, and natural materials such as stone or wood (Browning, Ryan and Clancy, 2014). Such individual (*I-focused*) space types include reflection/meditation rooms, music booths for sound baths, prayer rooms, nap pods, and outdoor reflection gardens. They exist to move users away from stress and into a tranquil setting where they feel transported away from the work environment without having to go far.

On the bottom right, **Create** spaces feature elements that promote freedom of thought and hands-on activities to enable an escape from technology and general office chaos. These together (*we-focused*) environments are playful zones for imagination and recreation. They provide needed stimulation via pops of bright colour, dynamic sound, and kinaesthetic or immersive technologies that excite the senses. If square footage is not available, designers can include "fidget features" like colouring walls, interactive light boxes, and simple puzzles that can be tucked away in smaller meeting or phone rooms for people to engage with when sitting in the space. A gallery area featuring employee artwork is another possibility. If larger square footage is available, consider options like music rooms where users can jam during breaks, and maker spaces for painting, sewing, or general creative tinkering. The goal in these spaces is to busy the hands and calm the mind.

Programming for revitalisation

To help determine the right space types for a facility, appropriate placement, and the number of revitalisation spaces needed, we recommend starting with the following questions.

Types of spaces

• Does the organisation's culture support this space type for revitalisation? Does leadership have a clear understanding of the benefits for offering these micro/macro-environments?

- Are there existing revitalisation spaces that should be enhanced with clear and authentic purpose for larger adoption and use?
- Would the creation of a revitalisation space type require an unexpected investment in new staff, equipment, and/or space that is not sustainable long term?
- Are there a variety of space types that can be accessed by all users and ability levels?
- Do any of the space types require adjacent storage needs that would impact the overall footprint?

Determining quantity and size

- How many users are expected to be in the space on a given day? Are there peak work days or shift workers that need to be considered?
- How far will they need to travel to access the space from their main work area?
- At what time of the day will the space be most utilised? If its peak utilisation is concentrated over a small period of time, a higher quantity may be necessary.
- Are there spaces that can be programmed to accommodate multiple uses, avoiding the allocation of additional square footage, but without causing operational or availability issues? If doubling up on uses, consider how much open floor space will be required for stretching or prayer after moving other items out of the way.

Determining location

- Are the spaces situated to ensure visual and acoustic privacy as needed for the space type?
- Are they located in areas where people feel comfortable using them and are reminded they exist? For example, avoid placing an entry off a corridor that only leads to the men's restroom. That may make the women in the office less likely to even know the space exists.
- Which spaces would benefit from access to daylight/views? Depending on the technology or lighting made available in a room to enhance an experience, designers may prefer a windowless space.
- Is there a need for the space to have plumbing access or specialty heating, ventilation, and air conditioning (HVAC)? How can you take advantage of existing locations to reduce the additional cost of these changes?

Design considerations – A journey to create revitalisation

Let's take a quick look at revitalisation in action followed by the design consideration checklist. Follow along in our workplace "day in the life" journey in the workplace and notice how these moments of wellbeing can be integrated throughout the entire day (Figure 14.4).

Figure 14.4 Revitalisation journey (graphics courtesy of IA Interior Architects and Lauren Newton).

Expand spaces

Recharge design considerations

- Snack areas/coffee zones provide a means to refuel the body for activity. Offer protein, carbs, and potassium-filled options that promote healthy choices.
- Include upbeat music to energise a workout zone. However, allow individual control of the acoustic environment where possible. Keep ambient music at a level that lets users listen to their own music via headphones without needing to blast the volume.
- Operable windows are ideal for fresh air and good ventilation especially when the space is crowded. Also, natural light helps keep circadian rhythms on track.

Moreover, for the display of edibles, daylight makes food appear at its most appealing and delicious.
- To maintain a healthy environment, high-touch areas require material selections that are pleasing to the senses and easy to sanitise.
- Consider providing equipment such as yoga rope walls or ladder walls that enable users to achieve a variety of physical attitudes and positions with support (i.e., inversions).

Connect design considerations

- The heart of a group activity area should be inviting to passers-by, but positioned where the action is neither visually nor acoustically distracting to those in nearby work or meeting zones.
- Music is often used to define and reinforce the brand or mood of a café or lounge area. In these areas, consider establishing zones with quieter noise levels for those who may wish to socialise but are easily overwhelmed by sound.
- The aroma and sight of freshly prepared, appetising food and drink is key when creating spaces that incorporate food. Be sure to separate these spaces from work zones so that aromas do not permeate the office.
- To create true comfort foods, consider leveraging the sense of taste by offering meals at the café from employee family recipes.
- Add outdoor amenities. Provide expansive environments and ensure relaxed expectations around acoustics for a connection with nature that creates a more comfortable and welcoming scenario where people can spend time together and build social relationships.
- Establish areas at the perimeter of communal spaces for refuge, for example, booths that allow those who feel more retiring to be part of the scene yet comfortable.
- Gaming areas can be relatively quiet (pool), relatively noisy (ping-pong), or highly variable (virtual reality or video games enjoyed with headphones or external speakers). In these interactive spaces, users should not be discouraged from having fun with one another in an animated way. Position gaming spaces in areas that do not distract from other functions and provide enough room for the intended number of occupants and activities.
- In virtual reality rooms where occupants are wearing headsets, safeguard measures should be in place to protect users from tripping or running into obstacles due to altered perceptions (visual, vestibular, proprioception).

Retreat spaces

Reflect design considerations

- Naturally quiet spaces are preferable to spaces that rely on sound masking to hide unwanted noise since white noise at typical office levels has proved to increase stress levels (Awada et al., 2022). Introducing sounds from nature such as running water or birdsong can stimulate our parasympathetic nervous system and give rise to a "rest and digest" response, which is how our bodies recover from stressful events.

- Provide fresh air whenever possible, and introduce delicate scents (such as lavender) to enhance the user experience of tuning in to the senses.
- Include access to items like tuning forks or singing bowls with a frequency of C or G which are the best-proven frequencies for relaxation, or access to other auditory sources like music with 60 BPMs or less (Magsamen and Ross, 2023). Curated binaural beats are another option; each ear hears a different frequency of less than 1000 Hz with a difference between them of no more than 30 Hz, which can induce relaxation and reduce anxiety (Smith, 2019.)
- The tactility of organic materials, how they feel to our hands or underfoot, offers a chance to connect with the natural world in a way that concrete, steel, and glass cannot provide.
- Visually, lighting is a powerful aspect of our experience of any space. Light enables us to perceive the objects within and boundaries of a space. The colour and intensity of light affect our circadian rhythms, our sense of time, and our ability to stay connected to natural rhythms of day and night. Provide colour tuning devices that allow a user to select the colour and temperature that best suits their mood.
- Provide options for resting positions – ensconced in an oversized armchair, curled up in a beanbag, or suspended in a slowly swaying hammock – to allow users to choose the kind of sensory input their nervous systems crave.

Create design considerations

- Activate stimulation via pops of bright colour, dynamic sound, and kinesthetic or immersive technologies that excite the senses.
- Design the entrance to feel welcoming for any creative skill level since research shows you need not be a master artist to reap the benefits of creating something with your hands (Magsamen and Ross, 2023).
- Activity should be invitingly visible to passers-by, but not visually or acoustically distracting to those in nearby work or meeting zones.
- These spaces can be quiet or active, depending on their function. Adjustable lighting, colour, access to exterior views, and sound systems are useful to dial in the right ambiance.
- Fidget features or interactive moments (e.g. colouring walls, interactive light displays, and other tactile elements) should be dynamic to avoid boredom in users.
- Music rooms, which may offer a combination of permanent equipment (amps, drum sets, keyboards) and instruments brought in by users, require careful planning and substantial soundproofing to ensure neighbouring areas are not negatively impacted.
- Gallery spaces that exhibit employee artworks or artworks from the community should be located in well-lit spaces on travel paths that are wide enough to let viewers stop and reflect without impacting the flow of traffic.

- Displays should be hung at a variety of heights for all to experience, whether from a seated position or from the vantage point of a person standing.

Design for a range of voices

Designing spaces that provide the right sensory experiences for restoration, renewal, and revitalisation while leveraging varying types and levels of participation will be key to supporting physiological and psychological health, as well as furthering inclusion, equity, and diversity. Studies show that stimuli need to fall along a continuum from quiet to highly stimulating, often correlating with introvert versus extrovert tendencies (DeYoung et al., 2014). Assessing individual differences through surveys or by hosting focus groups with end users can reveal optimal environments for revitalisation. Listening to a range of voices and observing work styles can help to address different sensorial needs and enable a fluid transition between inward and outward experiences, providing users with choice and more control over their preferred restorative experiences for health and wellbeing.

References

Abboushi, B., Elzeyadi, I., Taylor, R. & Sereno, M. (2019). Fractals in architecture: The visual interest, preference, and mood response to projected fractal light patterns in interior spaces. *Journal of Environmental Psychology, 61*, 57–70.

Abid, G., Zahra, I. & Ahmed, A. (2016). Promoting thriving at work and waning turnover intention: A relational perspective. *Future Business Journal, 2*(2), 127–137.

Adamson, K. & Thatcher, A. (2019). Do indoor plants improve performance outcomes?: Using the attention restoration theory: Volume VIII: Ergonomics and human factors in manufacturing, agriculture, building and construction, sustainable development and mining, *Proceedings of the 20th Congress of the International Ergonomics Association (IEA 2018)*, 591–604.

Aghniaey, S. & Lawrence, T. M. (2018). The impact of increased cooling setpoint temperature during demand response events on occupant thermal comfort in commercial buildings: A review. *Energy and Buildings, 173*, 19–27.

Allen, J. G., MacNaughton, P., Satish, U., Santanam, S., Vallarino, J. & Spengler, J. D. (2016). Associations of cognitive function scores with carbon dioxide, ventilation, and volatile organic compound exposures in office workers: A controlled exposure study of green and conventional office environments. *Environmental Health Perspectives, 124*(6), 805–812.

Altomonte, S., Allen, J., Bluyssen, P. M., Brager, G., Heschong, L., Loder, A., Schiavon, S., Veitch, J. A., Wang, L. & Wargocki, P. (2020). Ten questions concerning wellbeing in the built environment. *Building and Environment, 180*, 106949.

Andreucci, M. B., Loder, A., Brown, M. & Brajković, J. (2021). Exploring challenges and opportunities of biophilic urban design: Evidence from research and experimentation. *Sustainability, 13*(8), 4323.

Antonovsky, A. (1987). *Health Promoting Factors at Work: The Sense of Coherence* (Vol. 2022). Geneva: World Health Organization.

Aristizabal, S., Byun, K., Porter, P., Clements, N., Campanella, C., Li, L., Mullan, A., Ly, S., Senerat, A., Nenadic, I. Z., Browning, W. D., Loftness, V. & Bauer, B. (2021). Biophilic office design: Exploring the impact of a multisensory approach on human wellbeing. *Journal of Environmental Psychology, 77*, 101682.

Awada, M., Becerik-Gerber, B., Lucas, G. & Roll, S. (2022). Cognitive performance, creativity and stress levels of neurotypical young adults under different white noise levels. *Scientific Reports*, 12.

Bergefurt, L., Weijs-Perrée, M., Appel-Meulenbroek, R. & Arentze, T. (2022). The physical office workplace as a resource for mental health – A systematic scoping review. *Building and Environment, 207*, 108505.

Berto, R., Massaccesi, S. & Pasini, M. (2008). Do eye movements measured across high and low fascination photographs differ? Addressing Kaplan's fascination hypothesis. *Journal of Environmental Psychology, 28*(2), 185–191.

Boubekri, M., Cheung, I. N., Reid, K. J., Wang, C.-H. & Zee, P. C. (2014). Impact of windows and daylight exposure on overall health and sleep quality of office workers: A case-control pilot study. *Journal of Clinical Sleep Medicine, 10*(6), 603–611.

BRAG (2023). *Not at a Crossroads Anymore: ISSB, ESRS Ramp Up Disclosure Push.* BRAG. Retrieved Dec. 20 from https://br-ag.eu/2023/07/03/not-at-a-crossroads-anymore-issb-esrs-ramp-up-esg-disclosure-push/

Browning, W., Ryan, C. & Clancy, R. (2014). *14 Patterns of Biophilic Design: Improving Health & Wellbeing in the Built Environment.* New York: Terrapin Bright Green, LLC.

Bucher, T., Collins, C., Rollo, M. E., McCaffrey, T. A., De Vlieger, N., Van Der Bend, D., Truby, H. & Perez-Cueto, F. J. A. (2016). Nudging consumers towards healthier choices: A systematic review of positional influences on food choice. *British Journal of Nutrition, 115*(12), 2252–2263.

Butler, J. & Kern, M. L. (2016). The PERMA-Profiler: A brief multidimensional measure of flourishing. *International Journal of Wellbeing, 6*(3).

Capaldi, C. A., Dopko, R. L. & Zelenski, J. M. (2014). The relationship between nature connectedness and happiness: A meta-analysis. *Frontiers in Psychology, 5*(Aug.), 1–15.

Carrus, G., Scopelliti, M., Lafortezza, R., Colangelo, G., Ferrini, F., Salbitano, F., Agrimi, M., Portoghesi, L., Semenzato, P. & Sanesi, G. (2015). Go greener, feel better? The positive effects of biodiversity on the wellbeing of individuals visiting urban and peri-urban green areas. *Landscape and Urban Planning, 134*, 221–228.

Clements-Croome, D. (2017). Effects of the built environment on health and wellbeing. In *Creating the Productive Workplace.* Abingdon, Oxon: Routledge.

Clements-Croome, D., Turner, B. & Pallaris, K. (2019). Flourishing workplaces: A multisensory approach to design and POE. *Intelligent Buildings International, 11*(3–4), 131–144.

Colenberg, S., Jylhä, T. & Arkesteijn, M. (2021). The relationship between interior office space and employee health and wellbeing – a literature review. *Building Research & Information, 49*(3), 352–366.

Crandall, J. E., Space, I. & Space, G. (2018). The effect of light on interpersonal communication. *Journal of Environmental Psychology, 57*, 53–61.

de Dear, R., Xiong, J., Kim, J. & Cao, B. (2020). A review of adaptive thermal comfort research since 1998. *Energy and Buildings, 214*, 109893.

DeYoung, C. G., Quilty, L. C. & Peterson, J. B. (2014). Between facets and domains: 10 aspects of the Big Five. *Journal of Personality and Social Psychology, 107*(5), 880–896.

Dutcher, J. M., Lederman, J., Jain, M., Price, S., Kumar, A., Villalba, D. K., Tumminia, M. J., Doryab, A., Creswell, K. G., Riskin, E., Sefdigar, Y., Seo, W., Mankoff, J., Cohen, S., Dey, A. & Creswell, J. D. (2022). Lack of belonging predicts depressive symptomatology in college students. *Psychological Science, 33*(7), 1048–1067.

Dutton, J. E. & Ragins, B. R. (2006). *Exploring Positive Relationships at Work: Building a Theoretical and Research Foundation*. New York: Psychology Press.

Dzyuban, Y., Hondula, D. M., Vanos, J. K., Middel, A., Coseo, P. J., Kuras, E. R. & Redman, C. L. (2022). Evidence of alliesthesia during a neighborhood thermal walk in a hot and dry city. *Science of the Total Environment, 834*, 155294.

Ejlertsson, L., Heijbel, B., Brorsson, A. & Andersson, H. I. (2020). Is it possible to gain energy at work? A questionnaire study in primary health care. *Primary Health Care Research & Development, 21*, e65.

European Commission (2023). *EU Taxonomy for Sustainable Activities*. European Commission. Retrieved June 6 from https://finance.ec.europa.eu/sustainable-finance/tools-and-standards/eu-taxonomy-sustainable-activities_en

Fabritius, F. (2022). *The Brain-friendly Workplace*. Lanham, MD: Rowman & Littlefield Publishers.

Figueiro, M. G. (2017). Disruption of circadian rhythms by light during day and night. *Current Sleep Medicine Reports, 3*(2), 76–84.

Fonken, L. K. & Nelson, R. J. (2014). The effects of light at night on circadian clocks and metabolism. *Endocrine Reviews, 35*(4), 648–670.

Forooraghi, M., Cobaleda-Cordero, A. & Babapour Chafi, M. (2021). A healthy office and healthy employees: A longitudinal case study with a salutogenic perspective in the context of the physical office environment. *Building Research & Information, 50*(1–2), 134–151.

Fritz, C., Ellis, A. M., Demsky, C. A., Lin, B. C. & Guros, F. (2013). Embracing work breaks: Recovering from work stress. *Organizational Dynamics, 42*, 274–280.

Gatt, G. & Jiang, L. (2020). Can different types of non-territorial working satisfy employees' needs for autonomy and belongingness? Insights from self-determination theory. *Environment and Behavior, 53*(9), 953–986.

Gifford, R. (2014). *Environmental Psychology*, Fifth Edition. Colville, WA: Optimal Books.

Glazerman, G. (2019, June 10). Spotlight on Human Capital Has Investors' Attention. Value Reporting Foundation, SASB Standards. Retrieved Jan. 21 from https://www.sasb.org/blog/spotlight-on-human-capital-has-investors-attention/

Global Green Tag International (2023). *Nature Postive Standard+ and Nature Positive Declaration+*. Global Green Tag International. Retrieved Dec. 13 from https://www.global-greentag.com/npd-program.html

He, Y., Parkinson, T., Arens, E., Zhang, H., Li, N., Peng, J., Elson, J. & Maranville, C. (2022). Creating alliesthesia in cool environments using personal comfort systems. *Building and Environment, 209*, 108642.

Herneoja, A., Markkanen, P. & Juuti, E. (2022). An architectural viewpoint to user-centred work environment research to support spatial understanding in a transdisciplinary context through ecosystem-based approach. *Journal of Corporate Real Estate, 24*(3), 224–239.

Herway, J. (2017). *How to Create a Culture of Psychological Safety*. Gallup. Retrieved Nov. 22 from https://www.gallup.com/workplace/236198/create-culture-psychological-safety.aspx

Hongisto, V., Varjo, J., Oliva, D., Haapakangas, A. & Benway, E. (2017). Perception of water-based masking sounds – Long-term experiment in an open-plan office [Original Research]. *Frontiers in Psychology, 8*(1177).

International WELL Building Institute (2018). *The WELL Building Standard Version 2 Pilot (WELL v2)*. New York: Mind: International WELL Building Institute.

International WELL Building Institute (2022). *12 Competencies for Measuring Health & Wellbeing for Human and Social Capital*. International WELL Building Institute. Retrieved Dec. 20 from https://12competencies.wellcertified.com/

Jacques, J., Scholze, A., Galdino, M., Martins, J. & Ribeiro, B. (2018). Wellness room as a strategy to reduce occupational stress: Quasi-experimental study. *Revista Brasileira de Enfermagem, 71*, 483–489.

Jensen, P. A. & van der Voordt, T. J. M. (2020). Healthy workplaces: What we know and what else we need to know. *Journal of Corporate Real Estate, 22*(2), 95–112.

JLL (2023). *JLL Japan-Tokyo: HX Survey- Post Move Summary Analysis*.Chicago, IL.

Kaplan, R. (1993). The role of nature in the context of the workplace. *Landscape and Urban Planning, 26*, 193–201.

Kaplan, S. (1995). The restorative benefits of nature: Toward an integrative framework. *Journal of Environmental Psychology, 15*(3), 169–182. https://doi.org/10.1016/0272-49 44(95)90001-2

Kaplan, R. & Kaplan, S. (2005). Preference, restoration, and meaningful action in the context of nearby nature. In P. Barlett (Ed.), *Urban Place: Reconnecting with the Natural World* (pp. 330). Cambridge, MA: MIT Press.

Karasek, R. A. (1979). Job demands, job decision latitude, and mental strain: Implications for job redesign. *Administrative Science Quarterly, 24*(2), 285–308.

Kellert, S. R. (2018). *Nature by Design: The Practice of Biophilic Design*. New Haven, CT: Yale University Press.

Kellert, S. R. & Calabrese, E. F. (2015). *The Practice of Biophilic Design*. www. biophilic-design.com

Keyes, C. L. (2002). The mental health continuum: From languishing to flourishing in life. *Journal of Health and Social Behavior, 43*(2), 207–222.

Kim, J. & de Dear, R. (2013). Workspace satisfaction: The privacy-communication trade-off in open-plan offices. *Journal of Environmental Psychology, 36*, 18–26.

Konietzko, J., Das, A. & Bocken, N. (2023). Towards regenerative business models: A necessary shift? *Sustainable Production and Consumption, 38*, 372–388.

Koukounian, V. & Bourdeau, E. (2022, August). *The Role of Acoustical Privacy in a Hierarchical Framework for Acoustical Satisfaction: Past, Present, and Future Inter-noise.* International Institute of Noise Control Engineering (I-INCE), Wakefield, MA.

Kouvonen, A., Oksanen, T., Vahtera, J., Stafford, M., Wilkinson, R., Schneider, J., Väänänen, A., Virtanen, M., Cox, S. J., Pentti, J., Elovainio, M. & Kivimäki, M. (2008). Low workplace social capital as a predictor of depression: The Finnish Public Sector Study. *American Journal of Epidemiology, 167*(10), 1143–1151.

Lach, N., McDonald, S., Coleman, S., Touchie, M., Robinson, J., Morgan, G., Poland, B. & Jakubiec, A. (2022). Community wellbeing in the built environment: Towards a relational building assessment. *Cities & Health, 6*(6), 1193–1211.

Lahtinen, M., Ruohomäki, V., Haapakangas, A. & Reijula, K. (2015). Developmental needs of workplace design practices. *Intelligent Buildings International, 7*(4), 198–214.

Leesman Index (2020). *Home Working Survey*. Leesman Index. https://www.leesmanindex.com/

Li, K., Xu, S. & Fu, H. (2020). Work-break scheduling with real-time fatigue effect and recovery. *International Journal of Production Research, 58*(3), 689–702.

Licina, D., Wargocki, P., Pyke, C. & Altomonte, S. (2021). The future of IEQ in green building certifications. *Buildings and Cities, 2*(1), 907–927.

Lowe, G. S., Schellenberg, G. & Shannon, H. S. (2003). Correlates of employees' perceptions of a healthy work environment. *American Journal of Health Promotion, 17*(6), 390–399.

Lund, T., Labriola, M., Christensen, K. B., Bültmann, U., Villadsen, E. & Burr, H. (2005). Psychosocial work environment exposures as risk factors for long-term sickness absence among Danish employees: Results from DWECS/DREAM. *Journal of Occupational and Environmental Medicine, 47*(11), 1141–1147.

MacNaughton, P., Woo, M., Tinianov, B., Boubekri, M. & Satish, U. (2021). Economic implications of access to daylight and views in office buildings from improved productivity. *Journal of Applied Social Psychology, 51*(12), 1176–1183.

Madsen, I. E., Larsen, A. D., Thorsen, S. V., Pejtersen, J. H., Rugulies, R. & Sivertsen, B. (2016). Joint association of sleep problems and psychosocial working conditions with registered long-term sickness absence: A Danish cohort study. *Scandinavian Journal of Work, Environment & Health, 42*(4), 299–308.

Magsamen, S. A. & Ross, I. (2023). *Your Brain on Art: How the Arts Transform Us.* New York: Random House.

Marselle, M. R., Hartig, T., Cox, D. T. C., de Bell, S., Knapp, S., Lindley, S., Triguero-Mas, M., Böhning-Gaese, K., Braubach, M., Cook, P. A., de Vries, S., Heintz-Buschart, A., Hofmann, M., Irvine, K. N., Kabisch, N., Kolek, F., Kraemer, R., Markevych, I., Martens, D., … Bonn, A. (2021). Pathways linking biodiversity to human health: A conceptual framework. *Environment International, 150*, 106420.

Marzban, S., Candido, C., Avazpour, B., Mackey, M., Zhang, F., Engelen, L. & Tjondronegoro, D. (2023). The potential of high-performance workplaces for boosting worker productivity, health, and creativity: A comparison between WELL and non-WELL certified environments. *Building and Environment, 243*, 110708.

Marzban, S., Durakovic, I., Candido, C. & Mackey, M. (2021). Learning to work from home: Experience of Australian workers and organizational representatives during the first Covid-19 lockdowns. *Journal of Corporate Real Estate, 23*(3), 203–222.

Masento, N. A., Golightly, M., Field, D. T., Butler, L. T. & van Reekum, C. M. (2014). Effects of hydration status on cognitive performance and mood. *British Journal of Nutrition, 111*(10), 1841–1852.

McGee, B. & Marshall-Baker, A. (2015). Loving nature from the inside out: A biophilia matrix identification strategy for designers. *Health Environments Research & Design Journal, 8*(4), 115–130.

Mikus, J., Grant-Smith, D. & Rieger, J. (2023). Applying inclusive co-design and creative methods to elucidate a dual approach to eudaemonic home design: Crafting home and home crafting. In A. Maragiannis (Ed.), *Design in Action.* Elsevier.

Mishra, A. K., Loomans, M. G. L. C. & Hensen, J. L. M. (2016). Thermal comfort of heterogeneous and dynamic indoor conditions – An overview. *Building and Environment, 109*, 82–100.

Morgan, G. T., Coleman, S., Robinson, J. B., Touchie, M., Poland, B., Jakubiec, J. A. & … Cao, Y. (2022). Wellbeing as an emergent property of social practice. *Buildings and Cities, 3*(1), 756–781.

Morrison, R. L. & Smollan, R. K. (2020). Open plan office space? If you're going to do it, do it right: A fourteen-month longitudinal case study. *Applied Ergonomics, 82*, 102933.

Nurmi, N. & Pakarinen, S. (2023). Virtual meeting fatigue: Exploring the impact of virtual meetings on cognitive performance and active versus passive fatigue. *Journal of Occupational Health Psychology, 28*(6), 343–362.

Pachito, D. V., Eckeli, A. L., Desouky, A. S., Corbett, M. A., Partonen, T., Rajaratnam, S. M. & Riera, R. (2018). Workplace lighting for improving alertness and mood in daytime workers. *Cochrane Database System Review, 3*, CD012243.

Pálsdóttir, A. M., Spendrup, S., Mårtensson, L. & Wendin, K. (2021). Garden smellscape-experiences of plant scents in a nature-based intervention. *Frontiers in Psychology, 12*, 667957.

Plambech, T. & van den Bosch, C. C. K. (2015). The impact of nature on creativity – A study among Danish creative professionals. *Urban Forestry & Urban Greening, 14*(2), 255–263.

Policy Link (2012). Growing Urban Agriculture: Equitable Strategies and Policies for Improving Access to Healthy Food and Revitalizing Communities. Oakland, CA: Policy Link.

Rhoades, L. & Eisenberger, R. (2002). Perceived organizational support: A review of the literature. *Journal of Applied Psychology, 87*(4), 698–714.

Roskams, M., McNeely, E., Weziak-Bialowolska, D. & Bialowolski, P. (2021). Job demands-resources model: Its applicability to the workplace environment and human flourishing. In R. Appel-Meulenbroek & V. Danivska (Eds.), *A Handbook of Theories on Designing Alignment Between People and the Office Environment*. Abingdon, Oxon: Routledge.

Rozovsky, J. (2015). *The Five Keys to a Successful Google Team*. Google re:work. Retrieved Nov. 22 from https://rework.withgoogle.com/blog/five-keys-to-a-successful-google-team/

Schaufeli, W. B. & Taris, T. W. (2014). A critical review of the job demands-resources model: Implications for improving work and health. In *Bridging Occupational, Organizational and Public Health: A Transdisciplinary Approach* (pp. 43–68). Berlin, Germany: Springer Science + Business Media.

Smith, L. (2019). *What Are Binaural Beats, and How Do They Work?* Medical News Today. Retrieved Sept. from https://www.medicalnewstoday.com/articles/320019

Southon, G. E., Jorgensen, A., Dunnett, N., Hoyle, H. & Evans, K. L. (2018). Perceived species-richness in urban green spaces: Cues, accuracy and wellbeing impacts. *Landscape and Urban Planning, 172*, 1–10.

Spreitzer, G., Sutcliffe, K., Dutton, J., Sonenshein, S. & Grant, A. M. (2005). A socially embedded model of thriving at work. *Organization Science, 16*(5), 537–549.

Whelan, J., Sexton, K., Seeley, M., Nelson, M. L., Kashyap, M. & Bron, C. (2021). *Real Estate Strategy Reset: Eight Core Truths Guiding the Future of Work*. CBRE. Retrieved March 16 from https://www.cbre.com/insights/articles/real-estate-strategy-reset#introduction

Williams, K. J. H., Lee, K. E., Hartig, T., Sargent, L. D., Williams, N. S. G. & Johnson, K. A. (2018). Conceptualising creativity benefits of nature experience: Attention restoration and mind wandering as complementary processes. *Journal of Environmental Psychology, 59*, 36–45.

Wisdom Works Group (2023). Leading in the Health & Wellbeing Industry. Wisdom Works Group. Retrieved from https://www.bewellleadwell.com/wellbeing-leadership-study-report/

15 Spaces to encourage creativity

Editors' introduction

The creativity of people, working alone or in groups, contributes to the future success of an organisation. Creativity, for the purposes of this chapter, has been defined as developing and using original/imaginative ideas leading to innovation.

As the material in the pages that follow indicates, workplace design can facilitate and influence how people think creatively. But, of course, design alone does not enable people to think creatively, especially about topics with which they are not familiar.

Over generations, people have been creative in a wide range of spaces. These areas, whether artists' ateliers or chemistry labs or copywriter's workrooms, have provided the sort of relaxed but cognitively activated mindset that can support creative ideation. Design can make creative thinking more likely via surface colours, lighting, materials, soundscapes, scent profiles, furnishings, opportunities to influence the environment and architectural features.

DOI: 10.1201/9781003390848-15

Researcher perspective

Sally Augustin and Emily Dunn

Introduction

Fostering and promoting creativity and innovation within the workplace is a complex journey, one that acknowledges the multifaceted nature of individual and organisational needs. Design that does so recognises the need to support fundamental human needs and incorporates well-researched design solutions that take a human-centred approach.

We know that each person comes with their own unique personality types and their own set of tasks and objectives that may vary from day to day or even hour to hour. We need to consider a person's mood with respect to enhancing creativity and innovation as being in a slightly energised, more positive mood has been tied to more creative/innovative thinking (Isen et al., 1985; Isen et al., 1987; Cote, 1999; Grawitch et al., 2003; Baas et al., 2008; Byron et al., 2010; Hennessey and Amabile, 2010). Sander and colleagues (2019) and Veitch (2012) report on direct associations between environmental design, positive moods and enhanced creative/ innovative performance. A person's mood can be influenced by design (Desmet, 2015).

Researchers have thoroughly explored how workplace design can encourage people, both individuals and groups, to think more creatively/innovatively, consistently finding a link between the form of the workplace and creativity/innovation (Csikszentmihalyi, 1996; Sailer, 2011; Malinin, 2016).

Via objective research, investigators have determined that humans, individually and in groups, are likely to think more creatively/innovatively when they address the following factors and features.

General factors

- *Find themselves a biophilic designed environment* (Wijesooriya and Brambilla, 2021; Yin, 2021), for instance featuring natural materials (such as wood with visible grain and stone) and green leafy plants (McCoy and Evans, 2002; Shibata and Suzuki, 2002; Studente et al., 2016; Hall and Knuth, 2019; Stora Enso, 2020; Hahn et al., 2021). Seeing wood grain and thinking creatively/innovatively have been linked specifically (Stora Enso, 2020).
- *Are in places where environmental stressors/distractions have been minimised –* this includes spaces seeming appropriately spacious (Byron et al., 2010; Samani et al., 2015; Meinel et al., 2017; Thoring et al., 2019).

- *Can mentally refresh* (Gifford, 2014) *and socialise with colleagues* (Meinel et al., 2017).
- *Experience comfortable levels of environmental control* (about four to six options), they are not overwhelmed by the number of choices that can be made (Martens, 2011; Veitch, 2012; Samani et al., 2015).
- *Can customise/personalise their environments as individuals and teams* (McCoy, 2000; McCoy, 2002; Meniel et al., 2017).
- *Perceive particular sorts of messages in an area.* Positive nonverbal signals sent via design have been shown by Fong (2006) and Martens (2011) to be able to generally elevate creative/innovative performance; others have tied being able to read messages present as support for the activity-at-hand to enhanced creativity/innovation (McCoy, 2005; Dul and Ceylan, 2011; Dul and Ceylan, 2014; Thoring et al., 2019; Thoring et al., 2021).
- *Find that the design of the space needs to align with the activity-at-hand* (McCoy, 2005) – if a team needs to talk with each other, for example, they must be able to do so without distractions/interruptions.
- *Feel awed* (Yeung et al., 2011; Zhang et al., 2024). We are awed by a variety of different situations, for example, those where something is large, when rare or unusual materials are used, or exquisite workmanship is displayed, for example.

Visual features

- *See shades of the colour green* for even a short period (Lichtenfeld et al., 2012; Studente et al., 2016).
- *View colours that are unsaturated but light*; doing so has been tied to viewer mental states consistent with those linked to creativity/innovation noted above (Valdez and Mehrabian, 1994; Martens, 2011).
- *Experience moderate visual complexity* (McCoy and Evans, 2002; Ceylan et al., 2008; Vohs et al., 2013).
- *Are bathed in natural light* (Meinel et al., 2017). When *artificial light is in use,* creativity/ innovation gets a boost when it is *warmer*, about 3000 Kelvin, but not cooler, around 4500 or 6000 Kelvin (Weitbrecht et al., 2015; Abdullah et al., 2016) as well as *slightly dimmer* (for example, 150 versus 500 or 1500 lux) (Steidle and Werth, 2013).
- *Are in spaces with relatively more rounded physical design elements* (e.g. upholstery patterns and table shapes) than angular ones, those with sharper corners (Wu et al., 2021). The Wu-led team did find different levels of creativity/innovation, for example, when a tabletop in use was a curved shape than when it was rectangular.
- *Spend time in an area with windows providing views of nature* (McCoy and Evans, 2002; Ceylan et al., 2008; Dul and Ceylan, 2011; Loder and Smith, 2013, Van Rompay and Jol, 2016). The same effects are found when people can see indoor art depicting nature scenes (Batey et al., 2021).

Acoustic matters

- *Can hear nature soundscapes*, such as the pleasant noises present in a meadow on a lovely Spring day, burbling brooks, peacefully singing birds and gently rustling leaves and grasses (Browning and Walker, 2018).

Olfactory and indoor air quality factors

- *Can smell pleasant scents*, such as cinnamon-vanilla (Isen et al., 1997).
- *Are breathing cleaner air.* Arikrishnan et al. 2023 and colleagues found via a study conducted in a simulated office environment over 6 weeks that:

   higher TVOC [total volatile organic compounds] levels were significantly associated with lower-rated creative solutions. A 71.9% reduction in TVOC (from 1000 ppb), improves an individual's full creative potential by 11.5%. Thus, maintaining a low TVOC level will critically enhance creativity in offices.

Floor plan-related issues

- *Have the opportunity to move, walk around* (Oppezzo and Schwartz, 2014; Rominger et al., 2020; Murali and Handel, 2022). Oppezzo and Schwartz (2014), studying people walking (inside or outside, on a treadmill or not) found that walking's effects were immediate and briefly continuing, "Walking opens up the free flow of ideas, and it is a simple and robust solution to the goals of increasing creativity and increasing physical activity."
- *Are in a place with higher ceilings* (Meyers-Levy and Zhu, 2007). Meyers-Levy and Zhu collected data in spaces with eight- and ten-foot (2.4 and 3m) ceilings. Zhu and Mehta (2017) found that "when the room ceiling is perceived to be relatively high (versus low) it should enhance consumer creativity."
- *Are in a larger space.* We may also think creatively/innovatively in larger spaces. Chan and Nokes-Malach (2016) determined that "Larger room sizes facilitated the generation of more novel uses of everyday objects [divergent processing] … in Experiment 1; although the same trends were seen in Experiment 2, the effects did not reach significance." The researchers share that divergent problem-solving processes "have similar cognitive characteristics to the exploration stage of the creative process" and convergent ones are "processes that focus on 'converging' on a single 'correct' or canonical answer." Large rooms were 15'W × 30'L × 15'H (4.6 × 9.1 × 4.6 m) or 15'W × 30'L × 8'H (4.6 × 9.1 × 2.4 m) while small ones tested were 8'W × 10'L × 8'H (2.4 × 3 × 2.4 m).
- *Spend time in locations that increase the likelihood that they will communicate with people that they should interact with to think more creatively/innovatively* (Allen and Henn, 2007). Floor plans support creativity/innovation when the work areas of people who, if they spoke/emailed, would generate creative/innovative organisation-benefiting ideas are located adjacent to each

other, for example. Adjacencies might result in them sharing break areas, for instance.

- *Have access to private spaces*, when desired (Haner, 2005).
- *Can access a variety of different work areas*, as at different stages of the creative/innovative process, successfully completing different sorts of activities is key (Thoring et al., 2018). Changing from one work area to another has been tied to enhanced team creativity/innovation (Nicolai et al., 2016).
- *Are able to move to outdoor spaces.* As Palanica and colleagues (2019) report, "being outdoors in general may be enough to stimulate creativity, regardless of being surrounded by nature or a busy urban environment."
- *Are facing into the air current from the HVAC.* Izadi et al. (2019) found that: "frontal airflow (air blowing on the front of the body) boosts energetic activation and fuels enhanced performance on creative tasks, compared to dorsal airflow (air blowing on the back of the body)."

Furnishings and creativity

- *Are standing.* People may think more creatively/innovatively while standing (Baker et al., 2018). Not all are able bodied at any moment when creativity/innovation is required so walking and standing alternatives are important.
- *Are sitting on a cushioned surface* (Xie et al., 2016).
- *Have the opportunity to recline* (Michinov and Michinov, 2022).

Finally

Design can make it more likely that people will think creatively but it does not work alone. Organisational policies, training in relevant fields, e.g. marketing (if marketing slogans are being developed), have significant effects on creative performance.

Practitioner perspective

Craig Knight

Preamble

> Great minds discuss ideas, average minds discuss events, small minds discuss people.
>
> (Eleanor Roosevelt, 1901, cited in Skarmeas, 2014)

I am a psychologist. My days are filled with discussing people. My mind must be tiny. Although, without people, what ideas would there be, what events would showcase them and where would creativity come from? This chapter challenges Roosevelt absolutely. Ideas that do not lead to people are always suboptimal and frequently pointless. Nowhere is this concept more clearly illustrated than in the worlds of workspace design and business management, and underlined by the creativity they generate.

We'll start with design, which in its predictable top-down form produces a useable aesthetic, to which people's functions must adapt (Baldry and Hallier, 2010). The theocratic approach of the most influential workplace designers leaves little wiggle room. When asked about accepting input from those who use the workspace, Erik Veldhoen, the exemplary architect and developer of some of the most influential office space in the world, said "You can't just ask people what they want. You need professionals to lead the way." (Veldhoen, 2007). We need to query that statement.

If we are looking to develop a workspace where creativity flows – and in this chapter, we want nothing less – then, as we shall see, the job that designers and managers can do varies. The range runs from useful to the truly woeful.

What design-led solutions can *never* produce, however, is the *outstanding*. The Sterling Prize, Design Awards, BCO Office Awards and so forth are more about self-congratulation, about producing the salient, the on trend and impressing your peers; than they are about listening to the science and using small psychological minds to properly discuss the people. The people, however, are central to us, the psychology of their space is the crux of our chapter. Because without them, there can be no creativity.

We consider two key questions. First, what is it about office design that causes or prevents creativity? Second, what are the consequences of different strategies for creativity in the workplace?

Without correctly applied psychology, creativity in any space withers. Recent research has tied creativity in the workspace to freedom from management diktat, and trust in the workers to pursue their own routes to a job well done (Knight and Drummond, 2013)

The enriched workspace

Some organisations enrich their working environments with thinking spaces, break-out areas and discussion zones (McLennan, 2005). Design houses may variously recommend developing the office as a village (for finding resources), a street (for networking) or a library (as a knowledge centre, Duffy, 1997). Such initiatives are intended to enrich jobs, promote socialising and convenience and encourage creativity to flourish within a decent environment (Zelinsky, 2006). This contrasts sharply with the development of highly standardised, lean zones where corporate colours and instructional posters are often the only décor allowed (Kweon et al., 2008).

Figure 15.1 shows two similar spaces, but one is deliberately lean, with high-design chairs and desks, the other something of a dumping ground, accidentally enriched by plants and pictures. Consider the spaces, which feels the most comfortable?

Figure 15.1 Two similar but different spaces. (*source*: Knight)

Google has always tried to design enjoyable spaces, specifically to promote enjoyment of the job, innovation and creativity. Googleplex in California has been voted the best office in the world. It features napping pods, inter-floor slides, a piano in reception and a Tyrannosaurus Rex reminding employees not to become dinosaurs. Here, you will find two swimming pools and 18 cafes offering free food and drink (Entrepreneur Handbook, 2021).

There are many similar case studies of thoughtful design and workplace enrichment, which contrast sharply with modern lean, minimalist workspaces whose thick roots remain knotted in 19th-century working practices (Taylor, 1911). While advocates of lean offices recommend that décor should be restricted to pictorial communication of clear management goals, leaving workers in "no doubt about what the office is for" (Hobson, 2006), research that incorporates workspace enrichment has examined how pictures and plants can soothe mood, improve wellbeing and enhance creativity (Knight, 2017). Plants, for example, help to clean the air, but it is also argued that the simple act of enriching a space with artefacts not only makes that space more interesting and mentally stimulating, but also demonstrates managerial interest in the people that work there (Knight and Haslam, 2010a). This, in turn increases employees' feelings of belonging to the working community (Zeisel, 2006). Similarly, some organisations retain Bürolandschaft style screens to encourage movement across the working area and in this way increase social interaction (Zalesny and Farace, 1987). Others use screens simply to guide employees past a good view or towards a coffee point or café (Duffy, 1997; because social interaction – simply being with others and able to communicate – improves creativity (Haslam, 2004).

Enriching offices absorbs noise, lessens the sensation of overcrowding, reduces stress and increases feelings of privacy (Myerson, 2007), all of which are commendable improvements in any working environment. But it is the commercial benefits that are increasingly salient.

In practice

Managers or designers: Who has the answers to creativity

When we look to increase creativity in the workspace, the first obstacle to be overcome is existing practice. Given that science can find no benefits for lean workspaces, we should ask why do these uncreative, antediluvian spaces dominate the landscape? Our launchpad is the startling but ineluctable conclusion that business itself is stupid (Handy, 1990).

There is no excuse for lean practices. They are entirely toxic (Baldry et al., 1998). Blundering, aggressive, dull-minded business, snouting lean into as many crevices as it can manage, is doing actual damage (Knight, 2018).

No branch of any science on the planet thinks it is a good idea to keep animals in stark conditions, where they must conform to prescribed methods of action, under constant monitoring. Yet, this is precisely what lean conditions stipulate for human beings. Lean only exists as a good idea in business and business schools, it is risible outside the business bubble. There used to be lean care homes, as this made looking after people easier, but these were outlawed as cruel in 1948 National Assistance Board (1948).

All we need is a Google search – and a pair of high-grip waders to paddle through snake oil – to reveal considerable, impartial and easily accessible evidence that damns lean as worse than a con trick (Knight and Haslam, 2010a). Lean – or its sister discipline Six Sigma – will crucify creativity (Nieuwenhuis et al., 2014).

A lean workspace places the emphasis on money saved. Yet, when an office is enriched, the emphasis is placed on what is produced from what is spent. Looking after workers, rather than treating them as standard units of production, realises significant financial benefit through increased staff commitment and improved employee relations (Becker, 2004), it also improves all key output variables including creativity (Findler et al., 2007).

Lean and enriched office spaces: Similarities and differences

The first step to developing creative, productive spaces, therefore, is to improve the intelligence of the organisation. Every business is worth the effort, because of the people involved. And every business has the capacity to be exceedingly bright, first through harnessing the intelligence of its people and second by following the science. The huge stride that most businesses can take is simply to enrich the workplace.

There are many terms for enrichment from biophilic design, to the architecture of light, to hanging pictures, to having an on-site artist, and all of them are beneficial, and similarly effective. If you wouldn't put a zoo animal into human working conditions, don't expect people to work there either. Make space interesting. The basic rule is that the enrichment needs to be in the sight/smell/audio line of a workplace to make a difference (Nieuwenhuis et al., 2014).

There is no correlation between money spent and effects achieved. Installing a full biophilic building will be roughly as effective as filling a workplace with plants from the garden centre or hanging art on the walls (Knight and Haslam, 2010a). Indeed, the most creative spaces have been the cheapest to develop (Knight, 2018).

The key element in all enrichment programmes is love. The applied research indicates that if we show colleagues that thought and care have gone into the development of their environment, then the improvements in creativity and all key business variables has been in the tens of per cent compared to a lean workspace (Nieuwenhuis et al., 2014).

Anything that adds interest adds psychological engagement. Increasing psychological engagement is the fundamental building block of creativity. Meanwhile, the dull-brained business behemoth can be enticed with two irresistible titbits.

A Enriching a previously Spartan space offers more money for no extra effort at all.
B Management does not have to change its methods.

This is useful because the companies utilising lean methodology are likely to be particularly resistant to the structural and operational changes that would emanate from implementing wider changes (Braverman, 1974). Yet if psychology is to make a meaningful contribution to the lives of millions of working people, it is lean businesses that should be the main targets for collaboration.

Beyond enrichment

What of those organisations that are already there? Those that may not even have considered the hideous conditions of a lean space, preferring to look after their colleagues with gondolas in their canteens, beer in their breakout areas and objets d'art scattered about the workplace. Science is on their side. Creativity will be higher (Judkins, 2021). Is that it then? Is this global best practice?

Well, yes, according to industry experts. Again, do have a look at your favourite search engine. Ask it for the best office spaces in the world (there are some wonderful examples, from paddling pools in the working environment, to basketball courts by pod-shaped offices, to full-sized trees growing through the building's footplates) (e.g. Office Chai, 2017).

However, science shows us that global best practice is nothing special. The lauded spaces of Facebook in Menlo Park or Apple in Cupertino, achingly cool though they are, will help, but not maximise, creativity (Knight, 2017).

The most important ingredient for best results within any organisation is empowerment. So, give people a say over their workspace, preferably involving them as a team (Knight, 2018). Just as with the step up to enrichment, so moving onto empowerment further improves psychological engagement. This time the increase is linked to stronger senses of control and involvement (Knight and Haslam, 2010b). When teams create their own space, they have a vested interest in making that space work. All evidence points to a causal link between empowerment and improved performance, including enhanced creativity (Raymond and Cunliffe, 2000).

Too often organisations seem pressured into buying an "effective design solution" via badly informed consultancy, social pressure or fashion (Ryan et al., 2016). Business would do better to realise that it is more important to empower employees to customise their own workspace with familiar pictures and favourite plants than to provide them with anything that management has chosen no matter how tasteful and expensive (McLennan, 2005). Empowered workers feel happier in their workspace and are far more creative in the work they do (Knight, 2021a). There is also a useful inverse correlation between job satisfaction and job turnover (Elsbach and Bechky, 2007).

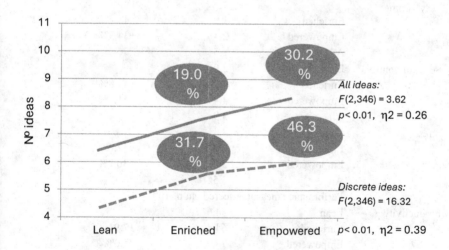

Figure 15.2 Ideas created in three environments (*source*: Knight and Drummond, 2013).

Research shows that creativity can increase by as much as 31% between lean and enriched conditions (Knight, 2021b), and by a further 13% between enriched and empowered (Knight and Drummond, 2013). Figure 15.2 shows how many ideas people developed in a simple creativity experiment. The experiment was interested in (a) how many ideas, on average, all participants developed and

(b) how many unique – or discrete - ideas (across teams of four) each person had. Note that in the Empowered condition, people had almost as many discrete ideas as participants in the Lean condition had ideas.

Table 15.1 shows the results across lean, enriched and empowered office conditions for a range of variables. Subjective variables are shown at the top of the table and are based on a seven-point scale (where the minimum score is 1, the maximum is 7). The objective variables productivity and creativity are scored out of 100. Note that all results here are in line with the science reported in this chapter (Knight, 2021b) and all differences are statistically significant.

Table 15.1 Results across lean, enriched and empowered office conditions

Variable	Condition	Mean	Change compared to lean
Job satisfaction	Lean	4.3	
	Enriched	5.6	+30.2%
	Empowered	5.5	+27.9%
	All	5.2	
Wellness	Lean	2.9	
	Enriched	5.2	+79.3%
	Empowered	5.3	+82.8%
	All	4.6	
Sick Office Syndrome	Lean	3.5	
	Enriched	1.9	−45.7%
	Empowered	2.3	−34.3%
	All	2.5	
Sense of control	Lean	2.9	
	Enriched	4.2	+44.8%
	Empowered	4.4	+51.7%
	All	3.9	
Stress	Lean	4.3	
	Enriched	2.9	−32.6%
	Empowered	3.0	−30.2%
	All	3.5	
Performance measures scored out of 100			
Productivity	Lean	54.6	
	Enriched	64.4	17.9%
	Empowered	73.7	35.0%
	All	64.3	
Creativity	Lean	44.1	
	Enriched	57.9	31.3%
	Empowered	64.9	47.2%
	All	56.0	

The data in this chapter meaningfully questions the lack of employee autonomy in the workspace and the gearing of space towards intensification, standardisation and control of work (McCarthy, 2020). The only function of lean practices within workspaces is the protection of managerial hegemony behind a façade of re-invented data that is becoming increasingly discredited (Baldry and Hallier, 2010). Replacing inequitable managerial practices with an empowered psychological approach to space management would put a business in great danger of benefitting *all* members of staff, boosting the bottom line and facilitating creativity.

Creative commercial advantages

Our chapter has advanced the idea that when organisations look to developing a creative space, they should consider and involve all the people who work within that space. Applying relatively straightforward psychological principles to the processes of space management can bring about significant and discernible changes to creativity at work. Empowering individuals and groups to have input into the design and structure of their own space delivers greater levels of not just creativity but of all other key organisational variables. So it is that creativity can act as a bellwether for a business (Reiter-Palmon and Hunter, 2023).

Figure 15.3 shows a model of the causal flow between management actions and creativity. Good design and empowerment within the workspace lead to a sense of feeling "at home" when at work (psychological comfort). The more at home people feel at work the more powerfully they identify with their employer. The higher the sense of identity, the happier people feel and the more creative they are. Note that the corollary is equally true (Knight and Haslam, 2010).

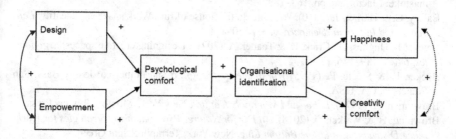

Figure 15.3 Model of flow between management actions and creativity.
Source: Knight and Haslam, 2010a.

It is time to develop an agenda for organisational change that roots out the injustices and poor creativity that have been at the heart of modern space management for over 100 years (Taylor, 1911). Only then can business effectively harness the creative power of its workforce.

So, to any organisation looking to foster creativity within the workspace, there are two fabulously counterintuitive rules to remember:

1 There is no such thing as the perfect space for creative working, yet...
2 Every organisation can have its own perfect space for creative working.

So for the best business solutions, forget Eleanor Roosevelt. Instead, discuss people at all times, put them at the heart of all your business ideas. Do more than listen to their input, allow them to develop the spaces in which *they* want to work. If that concession is made, then any organisation can develop communication, deepen commitment and drive creativity.

References

Abdullah, S., Czerwinski, M., Mark, G. & Johns, P. (2016). Shining (blue) light on creative ability. In *Proceedings, UbiComp'16, September 12–16*. Heidelberg: Association for Computing Machinery.
Allen, T. & Henn, G. (2007). *The Organization and Architecture of Innovation: Managing the Flow of Technology*. New York: Elsevier.
Arikrishnan, S., Roberts, A., Lau, W., Wan, M. & Ng, B. (2023). Experimental study on the impact of indoor air quality on creativity by serious brick play method. *Scientific Reports, 13*, 15488.
Baas, M., De Dreu, C. & Nijstad, B. (2008). A meta-analysis of 25 years of mood-creativity research: Hedonic tone, activation, or regulatory focus? *Psychological Bulletin, 134*(6), 779–806.
Baker, R., Coenen, P., Howie, E., Lee, J., Williams, A. & Straker, L. (2018). A detailed description of the short-term musculoskeletal and cognitive effects of prolonged standing for office computer work. *Ergonomics, 61*(7), 877–890.
Baldry, C., Bain, P. & Taylor, P. (1998). 'Bright satanic offices': Intensification, control and team Taylorism'. In P. Thompson & C. Warhurst (Eds.), *Workplaces of the Future*. Basingstoke: Macmillan, pp. 163–183.
Baldry, C. & Hallier, J. (2010). Welcome to the house of fun: Work space and identity. *Economic and Industrial Democracy, 31*, 150–171.
Batey, M., Hughes, D., Crick, L. & Toader, A. (2021). Designing creative spaces. *Ergonomics, 64*(1), 139–146.
Becker, F., & Steele, F., (1995) *Workplace by Design*, b San Francisco: Jossey Bass; San Francisco, CA, USA.
Braverman, H. (1974). *Labor and Monopoly Capital*. New York: Monthly Review Press.
Browning, B. & Walker, D. (2018). *An Ear for Nature: Psychoacoustic Strategies for Workplace Distractions and The Bottom Line*. New York: Terrapin Bright Green.
Byron, K., Khazanchi, S. & Nazarian, D. (2010). The relationship between stressors and creativity: A meta-analysis examining competing theoretical models. *Journal of Applied Psychology, 95*(1), 201–212.
Ceylan, C., Dul, J. & Aytac, S. (2008). Can the office environment stimulate a manager's creativity? *Human Factors and Ergonomics in Manufacturing, 18*(6), 589–602.
Chan, J. & Nokes-Malach, T. (2016). Situative creativity: Larger physical spaces facilitate thinking of novel uses for everyday objects. *Journal of Problem Solving, 9*, 29–45.
Cote, S. (1999). Affect and performance in organizational settings. *Current Directions in Psychological Science, 8*(2), 65–68.
Csikszentmihalyi, M. (1996). *Creativity*. New York: HarperCollins.
Desmet, P. (2015). Design for mood: Twenty activity-based opportunities to design for mood regulation. *International Journal of Design, 9*(2), 1–19.

Duffy, F. (1997). *The New Office*. London: Octopus.

Dul, J. & Ceylan, C. (2011). Work environments for employee creativity. *Ergonomics, 54*(1), 12–20.

Dul, J. & Ceylan, C. (2014). The impact of a creativity-supporting work environment on a firm's product innovation performance. *Journal of Product Innovation Management, 31*(6), 1254–1267.

Elsbach, K.D. & Bechky, B.A. (2007). It's more than a desk: Working smarter through leveraged office design. *California Management Review, 49*, 80–87.

Entrepreneur Handbook (2021). *10 Coolest Offices to Work In*. Retrieved from: https://entrepreneurhandbook.co.uk/10-coolest-offices-to-be-working-in/

Findler, L., Wind, L.H. & Barak, M.E.M. (2007). The challenge of workforce management in a global society: Modelling the relationship between diversity, inclusion, organizational culture, and employee wellbeing, job satisfaction and organizational commitment. *Administration in Social Work, 31*, 63–94.

Fong, C. (2006). The effects of emotional ambivalence on creativity. *The Academy of Management Journal, 49*(5), 1016–1030.

Gifford, R. (2014). *Environmental Psychology*, Fifth Edition. Colville: Optimal Books.

Grawitch, M., Munz, D., Elliott, E. & Mathis, A. (2003). Promoting creativity in temporary problem-solving groups: The effects of positive mood and autonomy in problem definition on idea-generating performance. *Group Dynamics: Theory, Research, and Practice, 73*, 200–213.

Hahn, N., Essah, E. & Blanusa, T. (2021). Biophilic design and office planting: A case study of effects on perceived health, wellbeing and performance metrics in the workplace. *Intelligent Buildings International, 13*(4), 241–260.

Hall, C. & Knuth, M. (2019). An update of the literature supporting the wellbeing benefits of plants: A review of the emotional and mental health benefits of plants. *Journal of Environmental Horticulture, 37*(1), 30–38.

Handy, C. (1990). *The Age of Unreason*. Boston: Harvard Business School Press.

Haner, U. (2005). Spaces for creativity and innovation in two established organizations. *Creativity and Innovation Management, 14*(3), 288–298.

Haslam, S.A. (2004). *Psychology in Organizations: The Social Identity Approach*. London: Sage.

Hennessey, B. & Amabile, T. (2010). Creativity. In S. Fiske, D. Schacter & C. Zahn-Waxler (Eds.), *Annual Review of Psychology*, vol. 53. Annual Reviews: Palo Alto, CA, 569–598.

Hobson, A. (2006). *The Obvious Office: Lean and the Obvious Office*. Norwich: Corvus Services.

Isen, A.M., Ashby, F.G. & Waldron, E. (1997). The sweet smell of success. *The Aroma Chology Review, VI*, 1, 4–5.

Isen, A., Daubman, K. & Nowicki, G. (1987). Positive affect facilitates creative problem solving. *Journal of Personality and Social Psychology, 52*(6), 1122–1131.

Isen, A., Johnson, M., Mertz, E. & Robinson, G. (1985). The influence of positive affect on the usualness of word associations. *Journal of Personality and Social Psychology, 48*(6), 1413–1426.

Izadi, A., Rudd, M. & Patrick, V. (2019). The way the wind blows: Direction of airflow energizes consumers and fuels creative engagement. *Journal of Retailing, 954*, 143–157.

Judkins, R. (2021). *Make Brilliant Work: Lessons on Creativity, Innovation, and Success*. London: Macmillan.

Knight, C.P. (2017). *Leadership, Led Design and Liberty: How to Create a World Class Workplace*. Helsinki, Client Report.

Knight, C.P. (2018). *New York Neighbourhoods: Lean versus Green.* New York, Client Report.

Knight, C.P. (2021a). *The End of the Affair, The Return to Offices and the Route to Happiness: Employment Lessons from the Pandemic.* London, Client report.

Knight, C.P. (2021b). *The Impact of Harvestable Plants on Performance and Wellbeing.* London, Client Report.

Knight, C.P. & Drummond, I. (2013). *Increasing Productivity Wellbeing and Creativity in Business: An Experimental Examination of the Psychological Application of Workspace.* London, Client report.

Knight, C.P. & Haslam, S.A. (2010a). Your place or mine? Organizational identification and comfort as mediators of relationships between the managerial control of workspace and employees' satisfaction and wellbeing. *British Journal of Management, 21,* 717–735.

Knight, C.P. & Haslam, S.A. (2010b). The relative merits of lean, enriched, and empowered Offices: An experimental examination of the impact of workspace management. *Journal of Experimental Psychology: Applied, 16,* 158–172.

Kweon, B.S., Ulrich, R.S., Walker, V.D. & Tassinary, L.G. (2008). Anger and stress - The role of landscape posters in an office setting. *Environment and Behavior, 40,* 355–381.

Lichtenfeld, S., Elliot, A., Maier, M. & Pekrun, R. (2012). Fertile green: Green facilitates creative Performance. *Personality and Social Psychology Bulletin, 38*(6), 784–797.

Loder, A. & Jerry Smith, J. (2013). Designing access to nature. *HealthCare Design, 13*(5), 58–63.

Malinin, L. (2016). Creative practices embodied, embedded, and enacted in architectural settings: Toward an ecological model of creativity. *Frontiers in Psychology, 6,* article 1978.

Martens, Y. (2011). Creative workplace: Instrumental and symbolic support for creativity. *Facilities, 29*(1/2), 63–79.

McCarthy, R. (2020). *Lean Methodology: A Guide to Lean Six Sigma, Agile Project Management, Scrum and Kanban for Beginners (Lean Thinking).* Independently published.

McCoy, J. (2000). *The Creative Work Environment: The Relationship of the Physical Environment and Creative Teamwork at a State Agency – A Case Study.* Doctoral dissertation. University of Wisconsin-Milwaukee.

McCoy, J. (2002). "Work Environments." In Robert Bechtel and Arza Churchman (eds.), *Handbook of Environmental Psychology,* John Wiley and Sons: New York, pp. 443–460.

McCoy, J. (2005). Linking the physical work environment to creative context. *Journal of Creative Behavior, 39*(3), 167–189.

McCoy, J. & Evans, G. (2002). The potential role of the physical environment in fostering creativity. *Creativity Research Journal, 14*(3–4), 409–426.

McLennan, J.F. (2005). *The Philosophy of Sustainable Design: A Journal.* Bainbridge Island: Ecotone.

Meinel, M., Maier, L., Wagner, T. & Voigt, K. (2017). Designing creativity-enhancing workspaces: A critical look at empirical evidence. *Journal of Technology and Innovation Management, 1*(1), 1–11.

Meyers-Levy, J. & Zhu, R. (2007). The influence of ceiling heights: The effect of priming on the type of processing people use. *Journal of Consumer Research, 34*(2), 174–186.

Michinov, N. & Michinov, E. (2022). The effects of postural feedback on divergent and convergent thinking. *Psychology of Aesthetics, Creativity and the Arts, 16*(3), 504–518.

Murali, S. & Handel, B. (2022). Motor restrictions impair divergent thinking during walking and during sitting. *Psychological Research, 86,* 2144–2157.

Myerson, J. (2007). Focus on: Office design. *Architects' Journal Specification, 1,* 23–28.

NationalAssistanceBoard.(1948).NationalAssistanceAct.Retrieved3rdMarch2022fromhttps:// www.legislation.gov.uk/ukpga/Geo6/11-12/29/enacted#:~:text=1948%20CHAPTER% 2029,%5B13th%20May%201948.%5D

Nicolai, C., Klooker, M., Panayotova, D., Husam, D. & Weinberg, U. (2016). Innovation in creative environments: Understanding and measuring the influence of spatial effects on design thinking-teams. In Hasso Plattner, Christoph Meinel, & Larry Leifer, (Eds.), *Design Thinking Research*. Switzerland: Springer International Publishing, pp. 125–139.

Nieuwenhuis, M., Knight, C.P., Postmes, T. & Haslam, S.A. (2014). The relative benefits of green versus lean office space: Three field experiments. *Journal of Experimental Psychology: Applied, 20*, 199–221.

Office Chai (2017). *These Are the 8 Best Offices in the World*. Retrieved from: https://office-chai.com/offices/best-offices-world/

Oppezzo, M. & Schwartz, D. (2014). Give your ideas some legs: The positive effect of walking on creative thinking. *Journal of Experimental Psychology: Learning, Memory, and Cognition, 40*(4), 1142–1152.

Palanica, A., Lyons, A., Cooper, M., Lee, A. & Fossat, Y. (2019). A comparison of nature and urban environments on creative thinking across different levels of reality. *Journal of Environmental Psychology, 63*, 44–51.

Raymond, S. & Cunliffe, R. (2000). *Tomorrow's Office: Creating Effective and Humane Interiors*. London: Taylor & Francis.

Reiter-Palmon, R. & Hunter, S. (2023). *Handbook of Organizational Creativity: Individual and Group Level Influences*. London: Academic Press.

Rominger, C., Fink, A., Weber, B., Papousek, I. & Schwerdtfeger, A. (2020). Everyday bodily movement is associated with creativity independently from active positive affect: A Bayesian mediation analysis approach. *Nature Scientific Reports, 10*, 11985.

Ryan, M.K., Haslam, S.A., Morgenroth, T., Rink, F., Stoker, J. & Peters, K. (2016). Getting on top of the glass cliff: Reviewing a decade of evidence, explanations, and impact. *Leadership Quarterly, 27*(3), 446–455.

Sailer, K. (2011). Creativity as social and spatial process. *Facilities, 29*(1/2), 6–18.

Samani, S., Rasid, S. & Bt Sofian, S. (2015). Individual control over the physical work environment to affect creativity. *Industrial Engineering and Management Systems, 14*(1), 94–103.

Sander, E., Caza, A. & Jordan, P. (2019). Psychological perceptions matter: Developing the reactions to the physical work environment scale. *Building and Environment, 148*, 338–347.

Shibata, S. & Suzuki, N. (2002). Effects of the foliage plant on task performance and mood. *Journal of Environmental Psychology, 22*(3), 265–272.

Skarmeas, N. (2014). *Quotations of Eleanor Roosevelt*. New York: Applewood Books.

Steidle, A. & Werth, L. (2013). Freedom from constraints: Darkness and dim illusion promote creativity. *Journal of Environmental Psychology, 35*, 67–80.

Stora Enso (2020). *10 Reasons Why Wooden Buildings Are Good for You and the Scientific Research to Back It Up*. Stockholm: Stora Enso. Retrieved from: https://info.storaenso. com/wood-house-effect.

Studente, S., Seppala, N. & Sadowska, N. (2016). Facilitating creative thinking in the classroom: Investigating the effects of plants and the colour green on visual and verbal creativity. *Thinking Skills and Creativity, 19*, 1–8.

Taylor, F.W. (1911). *Principles of Scientific Management*. New York: Harper.

Thoring, K., MilGoncalves, M., Mueller, R., Desmet, P. & Badke-Schaub, P. (2021). The architecture of creativity: Toward a causal theory of creative workspace design. *International Journal of Design, 15*(2), 17–36.

Thoring, K., Mueller, R., Badke-Schaub, P. & Desmet, P. (2019). An inventory of creative spaces: Innovative organizations and their workspace. In *Proceedings of the 22nd International Conference on Engineering Design,* August 5–8, 39048. Technical University Delft.

Thoring, K., Mueller, R., Desmet, P. & Badke-Schaub, P. (2018). Design principles for creative spaces. In *International Design Conference – Design 2018*. Glasgow, Scotland: The Design Society.

Valdez, P. & Mehrabian, A. (1994). Effects of color on emotions. *Journal of Experimental Psychology: General, 123*(4), 394–409.

van Rompay, T. & Jol, T. (2016). Wild and free: Unpredictability and spaciousness as predictors of creative performance. *Journal of Environmental Psychology, 48,* 140–148.

Veitch, J. (2012). Work environments. In S. Clayton (Ed.), *The Oxford Handbook of Environmental and Conservation Psychology*. New York: Oxford University Press, pp. 248–275.

Veldhoen, E. (2007). *Worktech Presentation*. London: Haworth Showrooms.

Vohs, K., Redden, J. & Rahinel, R. (2013). Physical order produces healthy choices, generosity, and conventionality, whereas disorder produces creativity. *Psychological Science, 24*(9), 1860–1867.

Weitbrecht, W., Barwolff, H., Lischke, A. & Junger, S. (2015). Effect of light color temperature on human concentration and creativity. *Fortschritte der Neurologie, Psychiatrie, 83*(6), 344–348.

Wijesooriya, N. & Brambilla, A. (2021). Bridging biophilic design and environmentally sustainable design: A critical review. *Journal of Cleaner Production, 283*, 124591.

Wu, Y., Lu, C., Yan, J., Chu, X., Wu, M. & Yang, Z. (2021). Rounded or angular? How the physical work environment in makerspaces influences makers' creativity. *Journal of Environmental Psychology, 73*, 101546.

Xie, J., Lu, Z., Ruiming, C. & Zhenguang, G. (2016). Remember hard but think softly: Metaphorical effects of hardness/softness on cognitive functions. *Frontiers in Psychology, 7*, 1343.

Yeung, W., Tschetter, A. & Shiota, M. (2011). Differential effects of distinct positive emotions on creativity. In APA Division 8, *Society for Personality and Social Psychology Conference,* January, pp. 27–29.

Yin, J. (2021). *Bringing Nature Indoors with Virtual Reality: Human Responses to Biophilic Design in Buildings*. Dissertation Abstracts International: Section B: The Sciences and Engineering, 82(6-B).

Zalesny, M.D. & Farace, R. (1987). Traditional versus open offices: A comparison of socio-technical, social relations and social relations and symbolic meaning perspectives. *The Academy of Management Journal, 30,* 240–259.

Zeisel, J. (2006). *Inquiry by Design*. New York: Newton.

Zelinsky, M. (2006). *The Inspired Workspace: Design for Creativity and Productivity.* Beverly: Rockport Publishers Inc.

Zhang, J.W., Howell, R.T., Razavi, P., Shaban-Azad, H., Chai, W.J., Ramis, T., Mello, Z., Anderson, C.L., Monroy, M. & Keltner, D. (2024). Awe is associated with creative personality, convergent creativity, and everyday creativity. *Psychology of Aesthetics, Creativity, and the Arts, 18*(2), 209–221.

Zhu, R. & Mehta, R. (2017). Sensory experiences and consumer creativity. *Journal of the Association for Consumer Research, 2*(4), 472–484.

16 To work!

Sally Augustin and Nigel Oseland

Overlapping solutions

The chapters in this book are arranged to address either design elements (such as lighting and acoustics) or how to create workspaces that support key work-related activities (like collaboration and individual focused work). But, of course, there are overlaps between the workplace solutions that address the topic of each chapter. This section highlights workplace solutions relevant to several topics.

Biophilia and cognitive performance

Chapter 2 covered biophilia, Chapter 13 was on focused work and Chapter 15 on creativity. The relationship between biophilic design, cognitive performance and creativity is well documented. For example, multiple studies over many years have linked being able to see green leafy plants, to higher levels of wellbeing and cognitive performance, especially creativity (Hall and Knuth, 2019; Nieuwenhuis et al., 2014; Studente et al., 2016). Natural materials, such as wood grain, can reduce stress levels and elevate cognitive performance (Poirier and Demers, 2019; Burnard and Kutnar, 2020). Furthermore, organic elements, such as rounded, as opposed to more angular physical environments and design elements, can result in higher levels of divergent creativity (Wu et al., 2021).

Areas flooded with natural light can enhance mood as well as cognitive performance (Edwards and Torcellini, 2002). To help overcome lack of natural light, circadian electric lighting or its colour and tone can help. Circadian lighting, to support and supplement that natural light, was found to align body rhythms with location, which is important for mood and mental performance (Beute and de Kort, 2014), as well as our health, as mentioned in Chapter 2. Tuning the colour and intensity of electric lighting in ways consistent with our evolutionary origins and biophilic design has important implications for our mindset and mental functioning (Weitbrecht et al., 2015).

Provide longer sight lines in some areas, opportunities for views with prospect and refuge in others, along with moderate levels of visual complexity. In areas with prospect and refuge, users feel secure and also have a view out over the nearby area. Heerwagen and Gregory's (2008) work supports these sorts of options for surveying the area-in-use, for example. Moderate levels of visual

DOI: 10.1201/9781003390848-16

complexity/clutter have also been tied to enhanced professional performance (Renner, 2020).

Consider providing a small fire or a water feature, especially in an adjacent outdoor space. Joye (2007) identified that both elements, along with deflected vistas, are important components of biophilic design.

Outdoor fresh air usually has a lower concentration of carbon dioxide (CO_2) than indoor air. Allen et al. (2016) determined that when occupants experienced CO_2 levels typically found in indoor spaces, their cognitive performance was worse than at lower concentrations of CO_2. Summarising much of the available research while reporting on his team's work, Allen (2021) shared that:

> My team at Harvard recently published research on the health of several hundred office workers around the world for more than a year. We found that people performed better on cognition tests when the ventilation rate in their working environment was higher. When they were they were exposed to more outdoor air, they responded to questions more quickly and got more answers right. Our team reached a similar finding a few years back in a tightly controlled laboratory setting. In that study, people did notably better on cognitive tasks when carbon dioxide made up about 600 parts per million of the air they breathed than when it made up about 1,000 parts per million … In our new research, we observed the effect in real buildings globally. We also observed a reduction of worker performance even at indoor CO_2 levels that many researchers had previously assumed were perfectly fine.

Biophilia and restoration

Chapter 14 focused on revitalising, which has clear links to Chapter 2 on biophilic design. Regardless of how people choose to spend their break time, particular in-office experiences will mentally refresh them. For instance, being able to look at a few green leafy plants (Raanaas et al., 2011) or seeing (at least some) nature out of windows (which is more plausible in some locations than others) or in art, either paintings or photographs or sculptures help cognitive refreshment (Kaplan, 1995; van den Berg et al., 2003; Kim et al., 2010; Veitch, 2012; Berto, 2014; Gifford, 2014). Open views, when our view out is not blocked by tree branches, etc., can be particularly refreshing (Herzog et al., 2003) and seeing green roofs is a restoring experience (Lee et al., 2014).

Featuring landscape elements in the workplace, like those that naturally occur in nearby green spaces, can also boost restoration (Menatti et al., 2019). Including wood with visible grain in the workplace can help restoration (Fell, 2010) as can moderately complex fractal patterns like those found in nature (Joye et al., 2016).

Scenes with visible, gently moving water seem to be particularly refreshing (White et al., 2010). A water feature in the office was identified as an important component of biophilic design (Joye, 2007), as is hearing and seeing a fish tank

(Cracknell, 2012). Furthermore, softly playing nature sounds or soundscapes, reminiscent of deciduous meadows on spring days (for example, burbling brooks, gently rustling leaves and grasses and quietly calling birds), have been tied to lower stress levels (Benfield et al., 2014) and possibly elevated cognitive performance (van Hedger et al., 2019; Moodsonic, 2022).

Related to restoration is mood and research consistently shows a link between mood, awellbeing and physical health – as mental state is elevated, physical processes follow (Sternberg, 2009; Segerstrom and Sephton, 2010). Physical stressors, like those temperatures that are a little too high or too low, or distracting noise, etc., can have particularly deleterious effects on mood and wellbeing.

Research has begun to focus on links between burnout and design, and burnout has both mental and physical components. Maslach (2017) suggests that organisations counter burnout among their members by focusing on "workload, control, reward, community, fairness, and values" (all of which have been discussed in this book). Reward, fairness and supporting values have all been linked to non-verbal signalling by the organisation that the contributions made by workers to the organisation's success are respected, valued and recognised. Workplace design is a powerful way to send signals to users.

Visual aesthetics, light and colour

Colour relates to both Chapter 4 on visual aesthetics and Chapter 5 on lighting. Colour is a topic that is frequently debated in vision-focused discussions. Fortunately, neuroscience research can resolve many a rigorous debate on the best colour of various surfaces in the workplace.

Colour has three attributes: saturation, brightness/lightness and hue. Hues are groupings of colours based on sets of wavelengths of light, such as reds, blues, greens, yellows, etc. Cultures form associations to hues, which is one of the reasons why it is important to involve occupants in any design effort. For example, it is much more desirable to use a shade in a space that is linked by workers to being trustworthy or dependable than to do the reverse. In the United States, for instance, the colour blue is associated with trustworthiness and dependability (Aslam, 2006), which is one reason why it is featured in so many bank interiors.

Saturation and brightness/lightness drive emotional responses to colours (Valdez and Mehrabian, 1994). Saturation can be thought of as the purity of a colour, with less saturated shades seeming to have more grey in them. Brightness/lightness is a colour's position along a black-to-white continuum with brighter/lighter colours being closer to white. Multiple researchers, including Valdez and Mehrabian (1994), have determined that colours that are less saturated and brighter/lighter are more relaxing to view while those that are more saturated and darker are more energising to look at. In spaces where people need to concentrate and to work with others, less saturated brighter/lighter colours are useful.

There are several colours that have been tied to specific cognitive outcomes. Seeing reds, even briefly, has been associated with degraded analytical performance

(Elliot et al., 2007; Maier et al., 2015). Viewing greens, however, has been linked to enhanced creativity (Lichtenfeld et al., 2012; Studente et al., 2016).

Colours are commonly categorised as warm (for example, oranges and yellows) and cool (like blues or greens). Research shows that seeing warmer colours leads to perceptions that the temperature in a space is slightly warmer than it is, while cool colours have the opposite effect (Mehta et al., 2012). Looking at people when they are in front of warm-coloured surfaces has also been tied to thinking that those people have "warmer" personalities (Choi et al., 2016). People in spaces featuring cooler colours feel more powerful than individuals where warmer ones are more plentiful (Dubois and Mehta, 2012).

Colours are often used in combinations and patterns. At all scales, moderate visual complexity, as discussed in Chapter 2, is preferred. Patterns, combining various colours and shapes at different scales and with varying degrees of order, can be more or less biophilic, or seemingly drawn from nature, which has important ramifications for occupant comfort, wellbeing, and performance. However, as the material in Chapter 2 on biophilia also indicates, patterns that are geometric, featuring rectangles, squares, triangles (and only the occasional circle, for instance) also have their place – they are good options when efficient action, such as travel down a hallway, is desirable.

There is more to visual aesthetics than just colour and whether curving or straight lines are most plentiful in a space. For instance, Chapter 4 highlights that humans prefer symmetric scenes, patterns, etc. to asymmetric ones (Du Sautoy, 2008; Brielmann and Pelli, 2018). Generally, symmetry is linked to pleasure, sophistication and calming situations, while asymmetric options are energising to experience (Bajaj and Bond, 2018). Balance is also very important when creating more pleasurable views (Kumar, 2016).

Noise and soundscapes

Not all sound is considered a noise or a problem, for example, Chapters 2 and 6 highlight that nature sounds in a space can reenergise people elevating their wellbeing and performance. In addition to nature soundscapes, sound masking systems (e.g. generating white and pink noise) are sometimes used to manage noise distraction from colleagues. However, research on the long-term benefits of sound masking reveals mixed results (e.g. Haapakangas et al., 2011; Perham and Banbury, 2011; Rugg and Andrews, 2010). As discussed in Chapter 11, neurodiverse workers (especially those hypersensitive to sound) are often stressed by unwanted acoustic stimuli.

Creating a soundless workspace is nigh on impossible, and in nature, complete silence can indicate danger – a storm, fire or predator. Having a faint background buzz in a workplace, not sufficient to be distracting, may be a benefit in some situations. For example, it reminds workers that they are not alone and part of a team, but also it may motivate them and even stimulate extroverts to perform better (Oseland and Hodsman, 2020). When conducting different sorts of tasks, different levels of environmental energy are useful (Appel-Meulenbroek et al., 2020). So, a slight background hum may be desirable when people are doing an oft-repeated task, one that doesn't require a lot of concentration and a quieter sound level is better when focus is required.

Designing spaces for meetings and solo work

Chapter 12 on collaboration and Chapter 15 on creativity indicated the need for good quality meeting spaces. There has been much research on how to create a successful meeting space.

Group spaces are most useful when they sit four to eight people (Vischer, 2011). Table shape and chair placement in meeting rooms has been extensively researched. When seats are arranged roughly in a circle, more of those present participate in the discussion (Sundstrom and Sundstrom, 1986), but if you are simply sharing information, then classroom-style arrangements in rows are better. More interestingly, people tend to prefer to talk to others across the corner of a table and are more likely to form relationships with them after they do (Sommer, 1969). Sitting on either side of a piece of furniture, say a table, creates psychological distance between people just as it insures a physical one (Sundstrom and Sundstrom, 1986) which can be quite handy if difficult, challenging, and stress-inducing topics need to be discussed.

Everyone needs an eye contact break from time to time (Argyle and Dean, 1965), so it is a good idea to include a mid-table plant or objet d'art etc. to which people can gracefully divert their eyes when a momentary eye contact break is required and not seeming rude is a priority. Western Europeans are more positive about eye contact than East Asians (Akechi et al., 2013) so "eye-diverters," whatever form they take, can be particularly important in some locations.

Researchers have learned that all meeting participants need to be sitting in chairs whose legs are the same length or all are sitting on the floor, i.e. no one is literally being looked up at or down on as they speak, to ensure that the contributions of all speakers are equally valued and respected (Bertamini et al., 2013; Makhanova et al., 2017; Baranowski and Hecht, 2018). Ackerman et al. (2010) found that people sitting on even a slightly padded seat interacted more productively and positively than ones sitting on harder chairs without cushions.

If all attendees can comfortably stand, then standing meetings can make gatherings briefer without compromising the quality of discussion (Bluedom et al., 1999). Knight and Baer (2014) found that standing by all attendees during meetings can improve a group's performance on knowledge work-type tasks.

Finally, meeting spaces without prospect and refuge (as described in Chapter 2) will be used less frequently, when teams have options, than those that do (Rashid et al., 2006).

Even highly collaborative working environments require spaces for solo work to prepare for presentations and develop ideas, etc. without distraction. Indeed, Wohlers et al. (2019) determined that activity-based working environments need work areas that support undisturbed solo work as well as spaces for collaborative sessions. Merely having workplaces for undisturbed solo work onsite supported job satisfaction, emotion-based plans to stay employed at an organisation and improve vitality. Work areas supporting collaboration had the same implications, particularly if the form of these places supported achieving what users needed to get done. Vitality was defined as liveliness and willingness to make an effort. Having enough collaborative spaces in place such that they were not all in use at the same time was more beneficial than having just the required number.

Chapter 11 focused on neurodiversity. An enclosed space, the size of a small, two-person meeting room that is well shielded acoustically and visually, is useful for neurodiverse individuals who need to re-establish a productive state but also handy for other workers requiring quiet space for concentration or contemplation or restoration. The requirements for such as space include the following design features:

- close-fitting blinds if the space has a window, to block view into this space completely,
- a way to allow neurodiverse individuals to run in place or listen to particular sounds at high volume or to take whatever other steps are required in their case to return to a productive mindset,
- not be on the conference reservation system, not be labelled and be available to anyone – they might be used by neurotypical people who need a place to do work requiring concentration, and this removes any potential stigma associated with their use,
- located in approximately the same place on each floor so that they are easy to find if a person's "usual spot" is in use and they need an alternative,
- ideally lockable from inside, to prevent people from entering unannounced, in error, etc., but an override for the lock must be readily available for the facilities team,
- one, possibly two, rooms of this type on a work floor are likely to be sufficient.

Bringing it all together

To recap all of the important and useful information contained in the last 14 chapters would require more pages than we have at our disposal. Furthermore, what is most important and useful can require adjustment from project to project, for example the key ways that employees add value at one organisation may not be the same as those at another. Similarly, in the case of a workplace refurbishment project, existing lighting conditions may be abysmal while acoustical ones are exemplary. Leaving the applied world, researchers often can only vary a carefully curated set of conditions in their investigations, not all of them, restricting their findings. So, what is a well-intentioned researcher or practitioner to do?

In Chapter 1, we reviewed Augustin's (2009) 5 Cs for places where people work to their full potential. These spaces *Coordinate* with work tasks, *Comfort* the occupants, *Communicate*, *Challenge* occupants to develop in ways they find meaningful and *Continue* in use over time, as detailed below.

- *Coordinate* with the task-at-hand. People who are trying to focus need to be able to concentrate without distraction and eyestrain, for example. As the chapters before this one have indicated, wellbeing and performance are influenced by a tremendous number of sensory experiences, seen, heard, felt, smelled and (occasionally) tasted, as well as additional factors such as organisational and national cultures and occupant personality profiles, etc.
- *Comfort* occupants. When bodies and minds are comfortable, for example, because they're in a biophilicly designed space, wellbeing and performance are likely to be at higher levels than when they're not.
- *Communicate* directly via nonverbal signals to users that their contributions to organisational success are recognised, respected and valued, and by supporting dialogues, spoken and unspoken, between those present. Furniture design, architectural design and design-induced mood, among other factors, all influence who we choose to talk to and what we say when we do.
- *Challenge* occupants to develop in ways that they value. People generally want to grow in ways that make it more likely that they'll do their jobs more effectively, as discussed in more detail in the sections in the first chapter on self-determination theory.
- *Continue* in use over time, evolving gradually, not shifting dramatically, but changing in psychologically manageable stages.

Oseland (2022) integrated the scientific evidence with practical experiences to identify 12 workplace design principles relevant to human-centric office design. There is much consistency between Oseland's and Augustin's lists. As Oseland reports, evidence-based human-centric design should incorporate the following high-level recommendations.

- Provide *Choices* for how and where and when work is done.
- Facilitate *Collaboration*, which requires fostering trust and mingling, which design can facilitate.
- Sustain *Concentration* as "some roles and work tasks do not actually benefit from continuous collaboration" and "High proportions of work time continue to be spent carrying out detailed or complex tasks," hence the ongoing need for spaces where people can concentrate, whether those are zones, rooms, pods, phone booths with walls, a roof and a door, or some other design solution.
- Encourage *Creativity* using a number of different sorts of workplace elements to aid this crucial endeavour.
- Offer *Comfort* either using ambient environmental features, though difficult, or through Choice and Control.
- Give *Control* because people are not all the same (consider the chapters on neurodiversity and personality) and opportunities to fine-tune conditions are definitely desirable.

- Provide *Connectivity* to people online and in-person, and to shared systems and applications.
- Support *Confidentiality,* not necessarily by providing private offices for all, but everyone needs access to spaces with good visual and acoustic shielding.
- Nurture *Contemplation,* because authors of multiple chapters in this book would agree with Oseland that "we sometimes just need a place to chill, relax, reflect and contemplate in solitude. Areas are required for gathering our thoughts and reenergising."
- Provide *Care* as employee wellbeing is mandatory.
- Ensure *Co-location* because being in the same place at the same time, at least occasionally, is required for teams to do their best work.
- Manage the *Cost* of a workplace project and its operation – the potential economic benefits of good design can be quantified (although it is sometimes difficult) and weighed against costs, as discussed in Chapter 1 and throughout *The Science of People and Office Design.*

Integrating the science contained in the last few hundred pages (and the many studies that were excluded for brevity), the insights provided by design practitioners, and the two C models of Augustin and Oseland, make it clear that effective workplaces need to meet the following criteria.

1 *Align with planned activities* – which may be more intellectual (e.g. coming up with new advertising slogans), social (like bonding with colleagues or developing team cohesion) or physical (for example, encouraging healthy choices). Developing a space where the things that need to happen can indeed take place involves lighting, ventilation, visual aesthetics and haptic experiences … the list is long. When alignment occurs, however, it supports Augustin's Coordination goal and Oseland's human-centric office requirements for Collaboration, Concentration, Creativity and Co-location.

2 *Build in opportunities for cognitive refreshment* – working minds get tired, particularly when they're doing focused work, alone or with others. Cognitive refreshment, as discussed in Chapter 14 on revitalising, is so crucial for individual and organisational wellbeing that it should be a planned activity as above. Workspaces that help occupants refresh cognitively are not only crucial to "getting the job done" but also support Oseland's objective of enabling Contemplation.

3 *Provide choice and environmental control* – including a variety of spaces with different environmental conditions in place. Humans do not all process incoming sensory information in the same way, and there is an assortment of factors, such as culture, that causes them to differentially utilise the material they take in. Choices and control let occupants select conditions that will work best for them as they work, all of which, as noted in earlier chapters, is good for wellbeing – mental and physical health. Workplaces that provide choice and environmental control are consistent with Oseland's human-centric office requirements for Choice, Control and Confidentiality.

4 *Adopt biophilic design principles* – Chapter 2 makes it clear that modern day Homo sapiens are not so different to the earliest members of the species in terms of how they process and choose to use incoming sensory information. Biophilic design acknowledges this consistency and supports both Augustin's and Oseland's Comfort goal, along with Oseland's criteria for Care.

5 *Send the right message nonverbally* – spending time researching and actually developing workplaces makes it clear that interpretations of situations can have a more powerful effect on how people think and behave than objective assessments (which may or may not be possible) of them. Culture(s) can have particularly strong effects on interpretation. So can an option that seems like it's environmentally responsible. Workspaces that communicate effectively are consistent with Augustin's Communication objective.

6 *Seem familiar* – familiar does not mean identical for now and forever, but evolving as everything must at a controlled pace, while always remaining understandable to users – no abrupt changes that leave people wondering how they can possibly utilise options presented to get their work done (let alone excel). When spaces are familiar, they continue, as Augustin requires.

7 *Take care with standardisation and automation* – not considering specific workplace requirements as work areas are developed, particularly ignoring occupant input, means creating a space that will not work and will most likely need to be modified again and again in the future, which is not sustainable. Workspaces that are cautious of over-automation and standardisation are consistent with Oseland's Connectivity and Cost goals.

8 *Recognise that not all humans are alike* – for starters, they may have different personalities, be neurodiverse and may be part of different (national and organisational) cultures. Individuals can differ from each other, but they tend to do so in predictable ways. For example, Chapter 10 indicated that people with similar personalities are attracted to similar jobs. These patterns make designing shared spaces much more feasible. Workspaces that acknowledge human differences are not only consistent with Augustin's and Oseland's Comfort factors, but Augustin's Challenge one as well.

So, valiant workplace researcher, design practitioner, researcher-practitioner/practitioner-researcher, remain dedicated, determined and returned to your professional endeavours. Use the material in this book to improve the professional lives and performance of workers and the organisations that employ them, today and tomorrow, and to make sure those current and future lives are truly worth living.

References

Ackerman, J., Nocera, C. & Bargh, J. (2010). Incidental haptic sensations influence social judgments and decisions. *Science, 328*(5986), 1712–1715.

Akechi, H., Senju, A., Uibo, H., Kikuchi, Y., Hasegawa, T. & Hietanen, J. (2013). Attention to eye contact in the west and east: Autonomic responses and evaluative ratings. *PLoS ONE,* 8(3): e59312.

Allen, J. (2021). Employers have been offering the wrong office amenities. *The Atlantic.* Retrieved from: https://www.theatlantic.com/ideas/archive/2021/10/fresh-air-cool-new-office-amenity/620288/

Allen, J., MacNaughton, P., Satish, U., Santanam, S., Vallarino, J. & Spengler, J. (2016). Associations of cognitive function scores with carbon dioxide, ventilation, and volatile organic compound exposures in office workers: A controlled exposure study of green and conventional office environments. *Environmental Health Perspectives, 124*(6), 805–812.

Appel-Meulenbroek, R., Le Blanc, P. & de Kort, Y. (2020). Person-environment fit: Optimizing the physical work environment. In O. Ayoko & N. Ashkanasy (eds.), *Organizational Behaviour and the Physical Environment.* New York: Routledge, pp. 251–267.

Argyle, M. & Dean, J. (1965). Eye contact, distance and affiliation. *Sociometry, 28*, 289–304.

Aslam, M. (2006). Are you selling the right colour? A cross-cultural review of colour as a marketing cue. *Journal of Marketing Communications, 12*(2), 15–30.

Augustin, A. (2009). *Place Advantage.* Hoboken: Wiley.

Bajaj, A. & Bond, S. (2018). Beyond beauty: Design symmetry and brand personality. *Journal of Consumer Psychology, 28*(1), 77–98.

Baranowski, A. & Hecht, H. (2018). Effect of camera angle on perception of trust and attractiveness. *Empirical Studies of the Arts, 36*(1), 90–100.

Benfield, J., Taff, B., Newman, P. & Smyth, J. (2014). Natural sound facilitates mood recovery. *Ecopsychology, 6*(3), 183–188.

Bertamini, M., Byrne, C. & Bennett, K. (2013). Attractiveness is influenced by the relationship between postures of the view and the viewed person. *I-Perception, 4*(3), 170–179.

Berto, R. (2014). The role of nature in coping with psycho-physiological stress: A literature review on restorativeness. *Behavioral Sciences, 4*, 394–409.

Beute, F. & de Kort, Y. (2014). Natural resistance: Exposure to nature and self-regulation, mood, and physiology after ego-depletion. *Journal of Environmental Psychology, 40*, 167–178.

Bluedorn, A., Turban, D. & Lore, M. (1999). The effect of stand-up or sit-down meeting formats on meeting outcomes. *Journal of Applied Psychology, 84*, 277–285.

Brielmann, A. & Pelli, D. (2018). Aesthetics. *Current Biology, 28*, R859–R863.

Burnard, M. & Kutnar, A. (2020). Human stress responses in office-like environments with wood furniture. *Building Research and Information, 48*(3), 316–330.

Choi, J., Chang, Y., Lee, K. & Chang, J. (2016). The effect of perceived warmth on positive judgment. *Journal of Consumer Marketing, 33*(4), 235–244.

Cracknell, D. (2012). The restorative potential of aquarium biodiversity. *Bulletin of People-Environment Studies, 39*, Autumn, 18–21.

Du Sautoy, M. (2008). *Symmetry: A Journey into the Patterns of Nature.* New York: HarperCollins.

Dubois, D. & Mehta, R. (2012). How warm and cool colors reverberate and shift psychological power. In *Proceedings of Annual Winter Conference, Society for Consumer Psychology*, February 16–18, Las Vegas, pp. 226–227.

Edwards, L. & Torcellini, P. (2002). *A Literature Review of the Effects of Natural Light on Building Occupants, NREL/TP-550-30756.* Golden, CO: National Renewable Energy Laboratory.

Elliot, A., Maier, M., Moller, A., Friedman, R. & Meinhardt, J. (2007). Color and psychological functioning: The effect of red on performance attainment. *Journal of Experimental Psychology: General, 136*(1), 154–168.

Gifford, R. (2014). *Environmental Psychology*, Fifth Edition. Colville: Optimal Books.

Hall, C. & Knuth, M. (2019). An update of the literature supporting the wellbeing benefits of plants: A review of the emotional and mental health benefits of plants. *Journal of Environmental Horticulture, 37*(1), 30–38.

Haapakangas, A., Kankkunen, E., Hongisto, V., Virjonen, P., Oliva, D. & Keskinen, E. (2011). Effects of five speech masking sounds on performance and acoustic satisfaction: Implications for open-plan offices. *Acustica, 97*(4), 641–655.

Herzog, T., Maguire, C. & Nebel, M. (2003). Assessing the restorative components of environments. *Journal of Environmental Psychology, 23*, 159–170.

Joye, Y. (2007). Architectural lessons from environmental psychology: The case of biophilic architecture. *Review of General Psychology, 11*(4), 305–328.

Joye, Y., Steg, L., Unal, A. & Pals, R. (2016). When complex is easy on the mind: Internal repetition of visual information in complex objects is a source of perceptual fluency. *Journal of Experimental Psychology: Human Perception and Performance, 42*(1), 103–114.

Kaplan, S. (1995). The restorative benefits of nature: Towards an integrative framework. *Journal of Environmental Psychology, 15*, 169–182.

Kim, T.-H., Jeong, G.-W., Baek, H.-S., Kim, G.-W., Sundaram, T., Kang, H.-K., Lee, S.-W., Kim, H.-J. & Song, J.-K. (2010). Human brain activation in response to visual stimulation with rural and urban scenery pictures: A functional magnetic resonance imaging study. *Science of the Total Environment, 408*, 2600–2607.

Knight, A. & Baer, M. (2014). Get up, stand up: The effects of a non-sedentary workspace on information elaboration and group performance. *Social Psychological and Personality Science, 5*(8), 910–917.

Kumar, M. (2016). Aesthetic principles of product form and cognitive appraisals." In Batra, R., Seifert, C. & Brei, D. (eds.) *The Psychology of Design: Creating Consumer Appeal.* New York: Routledge, pp. 234–251.

Lee, K., Williams, K., Sargent, L., Farrell, C. & Williams, N. (2014). Living roof preference is influenced by plant characteristics and diversity. *Landscape and Urban Planning, 122*, 52–159.

Lichtenfeld, S., Elliot, A.J., Maier, M.A. & Pekrun, R. (2012). Fertile green: Green facilitates creative performance. *Personality and Social Psychology Bulletin, 38*(6), 784–797.

Maier, M., Hill, R., Elliot, A. & Barton, R. (2015). Color in achievement contexts in humans. In A. Elliot, M. Fairchild & A. Franklin (eds.), *Handbook of Color Psychology.* New York: Cambridge University Press, pp. 569–584.

Makhanova, A., McNulty, J. & Maner, J. (2017). Love in the time of Facebook: Promoting attraction using strategic camera position. In *Proceedings of Personality and Social Psychology Conference*, January 21, San Antonio, TX, Program, p. 141.

Maslach, C. (2017). Finding solutions to the problem of burnout. *Consulting Psychology Journal: Practice and Research, 69*(2), 143–152.

Mehta, R., Chae, B., Zhu, R. & Soman, D. (2012). Warm or cool color? Exploring the effects of color on donation behavior. In *Proceedings of Annual Winter Conference, Society for Consumer Psychology*, February 16–18, Las Vegas, pp. 108–109.

Menatti, L., Subiza-Perez, M., Villalpando-Flores, A., Vozmediano, L. & San Juan, C. (2019). Place attachment and identification as predictors of expected landscape restorativeness. *Journal of Environmental Psychology, 63*, 36–43.

Moodsonic (2022). Nature-based immersion in the workplace. *Kinda Studios, London.* Retrieved from: https://www.moodsonic.com/news/nature-based-immersion-in-the-workplace-kinda-studios-report

Nieuwenhuis, M., Knight, C.P., Postmes, T. & Haslam, S.A. (2014). The relative benefits of green versus lean office space: Three field experiments. *Journal of Experimental Psychology: Applied, 20*, 199–221.

Oseland, N.A. (2022). *Beyond the Workplace Zoo: Humanising the Office.* Oxon: Routledge.

Oseland, N. & Hodsman, P. (2020). The response to noise distraction by different person-ality types: An extended psychoacoustics study. *Corporate Real Estate Journal, 9*(3), 215–233.

Perham, N. & Banbury, S.P. (2011). Do practical signal-to-noise ratios reduce the irrelevant sound effect? *Cognitive Technology, 16*, 1–10.

Poirier, G. & Demers, C. (2019). Wood perception in Daylit interior spaces: An experimen-tal study using scale models and questionnaires. *Bioresources, 14*(1), 1941–1968.

Raanaas, R., Evensen, K., Rich, D., Sjostrom, G. & Patil, G. (2011). Benefits of indoor plants on attention capacity in an office setting. *Journal of Environmental Psychology, 31*, 99–105.

Rashid, M., Kampschroer, K., Wineman, J. & Zimring, C. (2006). Spatial layout and face-to-face interaction in offices: A study of the mechanisms of spatial effects on face-to-face interaction. *Environment and Planning B: Planning and Design, 33*, 825–844.

Renner, R. (2020). *Why Pandemic Stress Breeds Clutter – and How to Break the Cycle*. National Geographic. Retrieved form: https://www.nationalgeographic.com/science/article/why-coronavirus-stress-breeds-clutter-how-to-break-cycle.

Rugg, M. & Andrews, M.A.W. (2010). How does background noise affect our concentra-tion? *Scientific American, January*. Retrieved from: https://www.scientificamerican.com/article/ask-the-brains-background-noise/

Segerstrom, S. & Sephton, S. (2010). Optimistic expectancies and cell-mediated immunity: The role of positive affect. *Psychological Science, 21*(3), 448–455.

Sommer, R. (1969). *Personal Space: The Behavioral Basis of Design*. Englewood Cliffs: Prentice-Hall.

Sternberg, E. (2009). *Healing Spaces: The Science of Place and Wellbeing*. Cambridge: Harvard University Press.

Studente, S., Seppala, N. & Sadowska, N. (2016). Facilitating creative thinking in the class-room: Investigating the effects of plants and the colour green on visual and verbal creativ-ity. *Thinking Skills and Creativity, 19*, 1–8.

Sundstrom, E. & Sundstrom, M. (1986). *Work Places: The Psychology of the Physical Envi-ronment in Offices and Factories*. New York: Cambridge University Press.

Valdez, P. & Mehrabian, A. (1994). Effects of color on emotions. *Journal of Experimental Psychology: General, 123*(4), 394–409.

van den Berg, A, Koole, S. & van der Wulp, N. (2003). Environmental preference and res-toration: (How) are they related? *Journal of Environmental Psychology, 23*, 135–146.

van Hedger, S., Nusbaum, H., Clohisy, L., Jaeggi, S., Buschkuehl, M. & Berman, M. (2019). Of cricket chirps and car horns: The effect of nature sounds on cognitive performance. *Psychonomic Bulletin and Review, 26*(2), 522–530.

Veitch, J. (2012). Work environments. In S. Clayton (ed.), *The Oxford Handbook of Environ-mental and Conservation Psychology*. New York: Oxford University Press, pp. 248–275.

Vischer, J. (2011). Human capital and the organization-accommodation relationship. In A. Burton-Jones, J.C. Spender & G. Becker (eds.), *The Oxford Handbook of Human Capital*. New York: Oxford University Press.

Weitbrecht, W., Barwolff, H., Lischke, A. & Junger, S. (2015). Effect of light color tem-perature on human concentration and creativity. *Fortschritte der Neurologie, Psychiatrie, 83*(6), 344–348.

White, M., Smith, A., Humphreys, K., Pal, S., Snelling, D. & Depledge, M. (2010). Blue space: The importance of water for preference, affect, and restorative ratings of natural and built scenes. *Journal of Environmental Psychology, 30*(4), 482–493.

Wohlers, C., Hartner-Tiefenthaler, M. & Hertel, G. (2019). The relation between activity-based work environments and office workers' job attitudes and vitality. *Environment and Behavior, 51*(2), 167–198.

Wu, Y., Lu, C., Yan, J., Chu, X., Wu, M. & Yang, Z. (2021). Rounded or angular? How the physical work environment in makerspaces influences makers' creativity. *Journal of Environmental Psychology, 73,* 101546.

Index

254   *Index*

sustainability 66; target melanopic EDI 66–67; textures, effect on 101; "three circles" model of lighting quality 63, *64*; TLM 67–68; workplaces 66–68, 73–74
Lister, K. 111
Lombard Effect 83
luminaires 68
Lynn, C. D. 9
Lyubomirsky 103

MacNaughton, P. 21
managers 229
Mankin, D. 168, 169
Marinez-Moyano, I. J. 168
Martin, J. 130
Maslin, S. 158, 160–162
Maslow's hierarchy of needs 152, *152*
mean radiant temperature (MRT) 26, 36
Meel, J. van 127, 128
meetings 176; and solo work, spaces for 243–244; standing 243
Meghan, T. 10
Meng, X. 47
mental health 7, 51, 152, 206
mental wellbeing 108
Merleau-Ponty, M. 95
metabolic rate 26, 36
Meyers-Levy, J. 225
mindfulness 55
Minkov, M. 121, 122, 124, 125
mobility 53
Mohd Jusan, M. 48
Moloughney, B. W. 106
motivation 2, 73, 105–108, 170, 189
multimedia hub 173
multi-sensory effects 12–13
Murphy, D. 106
Myers Briggs Type Indicators (MBTIs) 136–137, *137*, **139**
Myers, I. 136

Nadal, M. 45
Nagano, H. 96
Nanayakkara, K. 124
national culture 125; definition 121, 127–128; Hofstede's system 123; individualistic-collectivistic 121–122; individuality 123; indulgent-restrained 122–123; long-term orientation 123; power distance 123; time orientation 122; tolerance for uncertainty 122; toughness (masculine-feminine)

122, 123; uncertainty avoidance 123; uneven distribution of power, tolerance of 122
natural light 190–191, 194
nature sounds 13, 87, 203, 242
neurodiversity 149; acoustics and noise 159–160; auditory sense 153, 159–160; biophilic design 163–164; design for 156–164; features, design 244; gustation 154, 161; haptics 161–162; hypersensitive considerations 158; hyposensitive considerations 158; hypo- to hypersensitivity range *151*; indoor air quality 161; interoception 155, 161–162; lighting, colour and visual aesthetics 160–161; Maslow's hierarchy of needs 152, *152*; neurodistinctions 150–152; neurotypes 150; noise 159–160; odours 161; olfaction 153, 161; proprioception 154, 155; scope, design 157; sensory differences 158–159; sensory elements impact 152–155; tactile 154–155, 161–162; thermal comfort 161–162; ventilation 161; vestibular sense 155, 162–163; visual sense 153, 160–161
neurotypes 150
Newlands, A. 170
Newport, C. 187
Nicolai, C. 167
Nicoll, G. 107
Nocon, M. 106
noise 77, 82–83, 159–160; management 115; personality 137–138; and soundscapes 242–243
Nokes-Malach, T. 225
non-verbal communication 170
Non-visual Connection to Nature 17
No Place Like Utopia (Blake) 129
Nova, N. 170
nudges/nudging 105
Nummenmaa, L. 97
Nuwer 9

occupant personality *see* personality
OCEAN 136
Odbert, H. S. 135
odours *see* scents and odours
office acoustics 83–84; C$_{50}$ 84; soundscape 85–87
offices fundamentals: existence of work place 3–4; offices contribution to performance 5–6; performance metrics, design elements effect on **5**; task